Breeding

 BETWEEN

the Lines

Why Interracial People Are
Healthier and More Attractive

Alon Ziv

BARRICADE
BOOKS

Dedication

To my parents, who always told me I could do anything,
and then asked me why I hadn't yet.

Published by Barricade Books Inc.
185 Bridge Plaza North
Suite 308-A
Fort Lee, NJ 07024

www.barricadebooks.com

Library of Congress Cataloging-in-Publication Data
Ziv, Alon
Breeding between the lines: why interracial people are healthier and more
attractive/ Alon Ziv; with an introduction by Jay Phelan
p. em.
ISBN 1-56980-306-4 (pb.)
Racially mixed people. 2. Miscegenation. 3. Human genetics. I. Title.
HT1523.Z58 2006
305.8—dc22
2006042622

First Printing

Designed by India Amos, Neuwirth & Associates, Inc.
Manufactured in Canada

Contents

Acknowledgments

Alan Gordon earned this spot by winning the Name Alon's Book contest, but it's fitting that he be mentioned first. Alan is a tireless supporter and great friend and I relied on his unwavering enthusiasm at every step.

Krystal Houghton read every word and gave me extremely helpful feedback. But even more important than her editing, her unflagging excitement about the book reminded me why I was writing it.

Many friends contributed their time and support. Karen Gorodetzky and Joe Genier gave me the courage to start down this path. Tannaz Sassooni and Rachel Gandin read early drafts and provided fun, helpful notes.

My agent, Sharlene Martin, was a tireless and supportive advocate for the book.

I also want to thank: Bill Burger; Mike Dousette; Pammy Gordon, who has always been there for me; Heather Hammill and the entire Club Sandwich crew; Michelle Mehta; Deborah Sandy; Samantha Sher; Julia Vallone; David Viola, Edna Yaffe; Annie Yang; and all of the Zivs: Simon, Dalia, Eli, Rachel, Adina, and Ben.

Barricade Books deserves special thanks for making this book a reality and for taking risks in an industry that seems less and less likely to do so. I especially appreciate the efforts of Jeff Nordstedt and Carole Stuart.

Breeding Between the Lines is based on the research of countless scientists. Their relentless seeking of the truth made this book possible and has made the world a better place for all of us.

Finally, I can honestly say that *Breeding Between the Lines* would not exist if not for Jay Phelan. Jay's friendship, advice, and good humor have been invaluable over the years, but the most important thing he ever said to me was, "Of course you should write this book."

Foreword

ABOUT A HUNDRED and fifty years ago, Charles Darwin published *The Origin of Species,* smashing sacred ideas held by most of the world. Species could and did change over time and Darwin detailed the mechanism by which this occurred and was still occurring. The implications of his ideas extended far beyond the small community of scientists; they mattered to the general public. To this day, even in the midst of controversy and the continuing difficulties many have in accepting these truths, Charles Darwin's name and his ideas are known to virtually everyone.

The name Robert Bakewell does not elicit any such recognition. Yet nearly a hundred years before publication of *The Origin of Species,* Bakewell was already using the idea of evolution by natural selection. Bakewell revolutionized the practice of breeding animals, particularly the practice of selective breeding in order to improve particular features within a population. As the first to breed cattle for food, he made huge gains in improving their size and taste. Before Bakewell, the average bull weighed about 370 pounds. As a direct result of the techniques he developed, the average bull weight was more than doubled! He made similarly dramatic advances changing the very biology of sheep as well. None of his work was possible without an understanding of natural selection. But the world beyond practicing animal breeders continued on, contentedly holding ideas about the immutable nature of species that were absolutely inconsistent with the truths revealed in Bakewell's work.

This was not the first time that there was a disconnect between what non-scientists and scientists knew on an important issue, but it illustrates how dramatically out of step the worlds of scientists and non-scientists can be. In this case, the disconnect can probably be attributed to the fact that Robert Bakewell was interested less in the propagation of ideas and more in the propagation of animals. Whatever the cause, though, it mattered little to Bakewell that a formal, abstract description of what he was doing and why it could work did not exist. That was irrelevant. All that mattered was that it worked.

Today, there is a disconnect between scientists and non-scientists on the topic of race. On the one hand, the conventional wisdom among biologists is that race is not a useful concept. There is, they note, too much variation within any one "race," to make comparisons among different races. Yet throughout the world, race is acknowledged and used in a variety of ways, most notably on official government census forms and numerous large-scale public health studies. This book is an important step in addressing the disconnect.

Breeding Between the Lines is a dangerous book. Alon Ziv marshals the data that demonstrate the significant value of the concept of race in biomedical research. The data cannot be ignored simply because they are inconvenient for the current scientific dogma or because their political implications are divisive. But *Breeding Between the Lines* is a responsible book as well. Although he is not afraid to follow the data no matter where they lead, Alon also clearly addresses the limitations of the concept and takes a very even-handed approach. He has no agendas other than the truth and avoids the common pitfalls of overzealous analysis.

Writing about the biology of race requires appreciation of subtleties and critical nuances in the understanding of genetics, evolution, and physiology. In his work at UCLA, Alon Ziv has shown himself to be an accomplished educator in these disciplines, receiving accolades from his students not just for his deep understanding of the topics, but for his ability to generate and convey complex ideas in a focused and articulate way. Indeed, *Breeding Between the Lines* is that rare book that is at once insightful and revolutionary, while remaining compulsively readable and downright fun. We should be so lucky as to have a guide like Alon for all forays into deeply important but technically complex issues.

JAY PHELAN
Los Angeles, California
May 2006

1

Your Wedding Night

QUESTIONING WHAT OUR
PARENTS TAUGHT US

IMAGINE IT'S YOUR wedding night. The ceremony is over, and the guests have all gone home. You and your sweetheart are alone in a honeymoon suite filled with flowers and champagne. You're looking forward to a night of romance and passion with your soul mate—the person with whom you want to spend the rest of your life. You kiss, and as your eyes meet, you are filled with love and happiness. You kiss again. And again. As things start to heat up, you move over to the bed.

Now imagine the person slipping in between the sheets with you is your sibling.

How did your feelings change? Most likely, you started off feeling warm, romantic, and a little excited, but you ended up with a sick feeling in the pit of your stomach. The concept of incest is repulsive. We find it both disgusting and morally wrong.

Why do we have such strong and visceral feelings about incest? Our response is so deeply ingrained that we rarely question it, but for the answer we need look no further than television. Specifically to *The X-Files*. An episode from the fourth season titled "Home" revolves around the Peacocks, a family that has engaged in multiple generations of close incest. As a result, family members are grotesquely deformed, mentally deficient, and homicidally violent. The Peacock sons spend most of their time skulking in the shadows, bludgeoning people to death, and having sex with their crippled mother (on her orders). Viewers found the episode so disturbing it was never rerun on broadcast television. (If you are interested, it is available on video and DVD.)

The show is a supernatural thriller, a work of fiction designed to shock and titillate viewers, but it does highlight the chief reason we view incest as wrong: incest produces inferior children. They are more likely to suffer from genetic disorders, mental retardation, and early death. I'll discuss the reasons for this in Chapter Four.

Incest results in sickly offspring; therefore, we have evolved to avoid incest. This sounds reasonable, but what does it mean exactly? Geneticist Theodosius Dobzhansky once said, "Nothing in biology makes sense except in the light of evolution." In addition to having a cool name, Theodosius was a very smart man: to really understand biology—and this book—we have to understand evolution. Let's take a quick look at evolution and how it works.

THE EVOLUTION WILL NOT BE TELEVISED

In 1925, the Tennessee state legislature drew a line in the sand. They passed the Butler Act, which made it illegal to teach evolution in public schools. John Scopes soon crossed that line. A biology teacher in the small town of Dayton, Tennessee, Scopes continued to teach evolution and was promptly arrested for it. Sleepy Dayton became the setting of a national showdown between evolution and religion.

Defending Scopes, and consequently evolution, was the most famous trial lawyer in America, Clarence Darrow. His opponent was equally prestigious: William Jennings Bryan, who was known for his political career (a former secretary of state, he had unsuccessfully run for president three times) and his deep religious beliefs.

(Theodore Roosevelt once said of Bryan, "By George, he would make the greatest Baptist preacher on earth.") The stage was set for a spectacular face-off.

The trial was filled with great oratory and cheap theatrics. At one point, Bryan, who referred to evolution as "ape-ism," presented Darrow with a small wooden monkey. The courtroom was so crowded with eager spectators that the judge moved the trial to the courthouse lawn for fear the floor would collapse. The momentous trial lasted only eight days. In the end, Scopes was found guilty and fined $100. (His conviction was later overturned on a technicality.)

This much-hyped clash in Tennessee was only the first battle in the war over teaching evolution. As I write this, over eighty years later, some public schools in Georgia have placed warning stickers on textbooks that include evolutionary theory. The legality of the stickers will be decided in the courts, but clearly the debate between evolution and God rages on.

I was discussing this issue with my friend Pam one day when I had an interesting realization. It became clear to me that, in Pam's mind, evolution and God are not all that different! Pam strongly supports teaching evolution, but her ideas about what it is and how it works are somewhat vague. Pam is a very smart woman; she's just not familiar with the mechanisms behind evolution. She talks about it as a mysterious, almost magical, force that shapes animals perfectly and then guides them to their appropriate niche. It sounds . . . godlike!

The truth is that evolution is neither perfect nor magical. It's also not difficult to understand. It simply works by process of elimination. In school I used process of elimination on multiple-choice tests. I crossed out the answers that didn't seem as good until there was only one remaining. That final answer was the best fit. Evolution works the same way.

The cheetah's speed is a truly impressive evolutionary adaptation. But no guiding hand designed the cheetah to be fast. No magic here, just process of elimination. You can imagine that some cheetahs are faster than others. Slow cheetahs just aren't as good at catching gazelles. What happens to slow cheetahs? The savannah can be a cruel place—they starve to death. They die and don't leave offspring. Who's left in the population? Fast cheetahs. What kind of babies do fast cheetahs have? Fast babies.

Over time the slow cheetahs are weeded out, and the cheetah population gets faster and faster. This all happens through process of elimination. There's nothing magical or perfect about this process. If there were, cheetahs would have high-powered rifles. But even though evolution is a slow, incremental process, it can still produce impressive results—cheetahs can achieve a top speed of almost 70 mph.

As I was writing this section, I did a Web search for the word "cheetah," which returned 1.5 million results. That's a lot of Web pages, but I wasn't surprised since the cheetah is a fascinating animal. Just to put things in perspective, I did another search for "sex." This search returned over 90 million results! Apparently some things are far more interesting than cheetahs. This, too, makes sense in an evolutionary context.

We can imagine that ancient people had different ideas about sex. It's hard to imagine, but maybe some of them weren't that excited about sex. Maybe they found it boring. Maybe they were more interested in watching cheetahs. What happened to those people? They didn't have sex, so they didn't have babies. Asexual people are like slow cheetahs—they die out. Those ancient people who liked sex had babies, who had babies, who in turn had babies of their own, a process that eventually led to us. Those sex-loving cave people are our ancestors, and we have inherited their desires.

I mentioned previously that we have evolved to avoid incest. This just means that historically, individuals that mated with close relatives had inferior offspring and died out. Those who avoided mating with close relatives had healthy offspring and became our ancestors. If we want healthy, attractive children we must mate with nonrelatives. Our strong feelings of disgust toward incest ensure just that.

But what if avoiding incest is only part of the story? What if these lessons can be taken further? Perhaps the less "related" we are to our mates the better. Perhaps the more "different" you and your mate are the healthier and more attractive your children will be.

That's what this book is about. Mating outside our ethnic group or race can result in genetically healthier children. Children who are less likely to be sickly and more likely to be sexy.

To prove that interracial people have significant advantages, I'll delve into a wide variety of topics. I'll present scientific evidence distilled from academic journals in evolutionary biology, biochemistry, animal behavior, botany, anthropology, human diversity, and medical genetics. I've pored through thousands of human and animal studies so that you don't have to. I'll skip the academic jargon and give it to you straight.

I'll also discuss how race mixing has been viewed in the past in order to provide a historical perspective. And I'll analyze census data in order to get a clear picture of the current state of interracial relationships and interracial people. Finally I'll look to the future and make some predictions about the changing demographics of our increasingly integrated world. Plus I'll throw in some anecdotes and movie references just to keep things interesting.

My theory of interracial advantages flies in the face of the current party line. It has the rare ability to provoke both conservatives and liberals. Because of the controversial nature of the topic, it's critical that everything checks out. My tone may be light at times, but the information I'm presenting is serious. Every fact, be it a scientific study, historical account, or pop culture tidbit has been carefully researched. I encourage you to check out the notes section in the back of the book for references or more information on a particular topic.

This book is about taboos. We just discussed the biological taboo of incest, but when we move to the other end of the spectrum, we also run into serious controversy. Many of the battles of the civil rights movement have been fought and won, but the world is still far from color-blind. Mixed marriages remain taboo and frequently lead to conflict, even violence. Historically, race mixing has been socially and often legally prohibited. Mixed individuals were often dismissed as "mongrels."

Yet They Called Him the Georgia Peach

When legendary Detroit Tigers baseball player Ty Cobb was heckled by a fan in May 1912, he brushed it off. But when the heckler called him a "half-nigger," Cobb, who was white, immediately leapt over the railing, rushed up twelve rows of seats, and started beating the man savagely. The man was quickly knocked over by the vicious assault, but Cobb wasn't satisfied and began stomping on him. To make matters worse, Cobb was rumored to sharpen the metal spikes on his cleats with a file in order to be more intimidating when he slid into base. Cobb refused to stop his brutal attack even when an onlooker pointed out that the heckler had only one hand! Cobb's response—"I don't care if he has no feet." Not exactly Baseball's Ambassador of Goodwill.

Ty Cobb was notoriously racist and hot-tempered, but the most shocking detail of this story is its conclusion. After this violent display, Cobb was suspended indefinitely. In response, his teammates protested and refused to play until he was reinstated. It was one of the first effective players' strikes, and Cobb got off with only a ten-day suspension. But Cobb's teammates didn't rally around him because he was popular. In fact the opposite was true—his erratic behavior had made even his own team hate him. Two years earlier when it looked like Cleveland's Napoleon Lajoie was going to edge out Cobb for the American League batting championship, Lajoie received a telegram of congratulations—from eight of Cobb's teammates! So why did they stand up for Cobb? They supported him because they felt that the heckler's comment was so insulting that Cobb's response was reasonable.

They weren't the only ones who thought so. After the incident was over, the congressional delegation from Cobb's home state of Georgia commended him for his actions. They made it clear that he had acted as a proper son of the South in defending his honor against such a grievous insult.

This wasn't just the view of some kind of racist baseball fringe; it was the prevailing opinion of the day. Fifty years earlier, Paul Broca was studying race and intelligence in Paris. Broca was no fly-by-night scholar; he was an eminently respected physician and founder of the Anthropological Society of Paris. When I took neuroanatomy in college, I studied the part of the brain responsible for speech. It's name? Broca's area.

In addition to getting parts of the brain named after him, Dr. Broca was one of the experts of his day on the topic of race science. What was his learned opinion of people of mixed heritage? He proposed that biracial individuals were mentally, morally, and physically inferior to those of pure race.

In America a similar argument was presented by Louis Agassiz, a prestigious Swiss naturalist who founded and directed the Museum of Comparative Zoology at Harvard. Agassiz was far less subtle than Broca about his feelings toward miscegenation. He described "halfbreeds" as "a sin against nature" similar to incest. He went on to editorialize, "The idea of amalgamation is most repugnant to my feelings, I hold it to be a perversion of every natural sentiment."

This was a sneaky attempt to link a social taboo with a biological one. Agassiz was trying to tar intermarriage with the same brush as incest in order to make it seem unnatural and wrong. Ironically, incest and intermarriage are opposites and yield opposite results in terms of offspring.

But if race mixing was so unnatural and repulsive, why was it a common occurrence in the South? Agassiz actually proposed the ridiculous argument that innocent Southern boys were being seduced by hot-to-trot "colored house servants" who were insatiably drawn to their white masters.

At this time race scientists were focused on analyzing and ranking the major ethnic groups (usually divided into whites, blacks, Asians, and Native Americans). Somehow whites always seemed to come out on top, often because scientists consciously or subconsciously distorted the data. The status of biracial individuals was an important topic for two reasons. First, the conventional wisdom held that whites were superior; therefore they should remain separate so as not to be "diluted" by other races. Second, some, like Agassiz, believed that whites and blacks were actually two separate species and shouldn't be biologically able to interbreed successfully. These scientists supported this fallacy by claiming, without proof, that mixed people were mentally and physically unfit. (Obviously Agassiz never met Halle Berry.)

This thinking is also the origin of the term *mulatto*, which refers to a person of mixed black and white heritage. I was shocked to learn that this word is a reference to the mule. Mules are hybrids between horses and donkeys (which are two different species), and while they are good at transporting you to the bottom of the Grand Canyon, they are sterile and thus defective. I always felt bad that Native Americans were called "Indians" because of Columbus's mistaken impression of where he had landed, but it seems equally terrible to be named after a beast of burden. Can you imagine a life where people continually call you "donkey" to your face?

In Chapter Five, we'll see why mulatto is a misnomer. The mule analogy is a fallacy; biracial individuals are far from sterile. In fact, they are extremely fertile. I'll also tell you about other examples taken from agriculture that make more accurate comparisons.

BREAKING THE LAW

Race mixing wasn't just frowned upon socially and scientifically, but also legally. Prohibitions against interracial marriage date back to colonial times; there's even a name for them: antimiscegenation laws. The consequences for breaking these laws, while harsh, were also depressingly practical. Whites who broke the law were often driven from the colony; blacks, on the other hand, were far too valuable as slaves to banish.

Despite strict penalties (in colonial Virginia, a minister who performed an interracial wedding was fined ten-thousand pounds of tobacco) the laws were often broken covertly. The same slave owners who passed these laws often used their female slaves sexually, much of the time by force. Originally colonial law determined the status of a child (slave or free) by his father's status. The number of female slaves impregnated by their masters made this somewhat inconvenient. The solution? Legislators changed the law to follow the mother's status.

As early as 1691, the Colony of Virginia outlawed miscegenation. Other colonies (and later states) followed suit. The prevention of inferior offspring was often cited to legitimize these laws. A Georgia court in 1869 claimed, "The amalgamation of the races is not only unnatural, but is always productive of deplorable results." Nor was this only a Southern phenomenon. In 1878, Senator John F. Miller from California (which had attracted thousands of Chinese immigrants during the Gold Rush) said, "Were the Chinese to amalgamate at all with our people, it would be the lowest, most vile and degraded of our race, and the result of that amalgamation would be a hybrid of the most despicable, a mongrel of the most detestable that has ever afflicted the earth." His political hyperbole paid off. Two years later, California (which the 2000 census found to be home to more interracial couples than any other state) passed its own antimiscegenation law prohibiting intermarriage between whites and "mulattos," "Negroes," or "Mongolians" (East Asians).

These views did not mellow over time. In 1924, Virginia's Racial Integrity Act turned up the heat and made interracial marriage a felony punishable by one to five years in prison. Thirty-four years later, these laws were still on the books. It was in this legal climate that Mildred Jeter, a black woman, and Richard Loving, a white man, met and fell in love.

It was 1958, and interracial unions were illegal in sixteen states. Only three years earlier, the Virginia Supreme Court had upheld the validity of its antimiscegenation law saying its legitimate goal was "to preserve the racial integrity of its citizens" and prevent "the corruption of blood," "a mongrel breed of citizens," and "the obliteration of racial pride."

Since they couldn't get married in their native Virginia, Mildred and Richard crossed the state line and tied the knot in Washington, D.C. Shortly after, they returned to their home state and were soon arrested for violating the antimiscegenation law. The appropriately named Lovings pleaded guilty and were both sentenced to one year in prison. (Ironically the Virginia state motto is "Virginia is for Lovers.")

The judge was quite clear in his opinion, "Almighty God created the races white, black, yellow, malay and red, and he placed them on separate continents. And but for the interference with his arrangement there would be no cause for such marriages. The fact that he separated the races shows that he did not intend for the races to mix." After relaying God's will, the judge generously offered to suspend their sentences if the Lovings left Virginia and did not return for twenty-five years. While exile is certainly better than jail, it isn't much of a wedding present.

The Lovings moved to Washington D.C., but decided to fight for their right to be married in Virginia. It would be a difficult fight: only 4 percent of white Americans approved of mixed marriages. They filed a motion to set aside their sentences. Nothing happened. They instituted a class action lawsuit claiming Virginia's antimiscegenation law was unconstitutional. They lost. They appealed. And lost again. Finally, in 1967 their case reached the Supreme Court. The court unanimously sided with the Lovings and declared antimiscegenation laws unconstitutional as they violated the "equal protection" guaranteed by the Fourteenth Amendment. Nine years after their wedding, Richard and Mildred could live as man and wife in their home state of Virginia.

TECHNICOLOR LINES

Also in 1967, moviegoers had the opportunity to watch an all star cast explore these same issues in the drama, *Guess Who's Coming to Dinner?* Sidney Poitier and Katharine Houghton play an interracial couple who have fallen in love and now have to tell their parents. Spencer Tracy, who plays Katharine Houghton's father, puts it best, "You're two wonderful people who happened to fall in love and happen to have a pigmentation problem."

But is pigmentation still a problem? I'd like to believe that this is all ancient history. After all, *Guess Who's Coming to Dinner?* was recently remade as *Guess Who?*—a comedy starring Bernie Mac and Ashton Kutcher. It certainly says something that we can now approach these same themes with slapstick comedy and sexual innuendo.

This remake is no surprise—forty years later, Hollywood is still fascinated by interracial unions. Many movies have dealt with this highly charged topic. One of the most successful was 1992's *The Bodyguard* starring Whitney Houston as a pop diva and Kevin Costner as her loyal bodyguard. Their romance crossed race and class lines and, along with a great soundtrack, propelled the film to become an international blockbuster. But the idea for *The Bodyguard* wasn't new. The movie was originally proposed in the 1970s with Diana Ross opposite either Steve McQueen or Ryan O'Neal, but was shelved because it was deemed too controversial.

Also controversial was Spike Lee's *Jungle Fever*, the story of a successful black architect, played by Wesley Snipes, who has an affair with a white coworker played by Annabella Sciorra. The racial divide and the fact that he is married make for a very taboo relationship. The film gives a variety of perspectives on interracial dating and is surprisingly balanced, given Spike Lee's open disapproval of mixed couples. While promoting the movie he said, "I give interracial couples a look. Daggers. They get uncomfortable when they see me on the street."

Perhaps the prevalence of that attitude can help explain a confusing experience I had a couple of years later when I went to see the movie adaptation of John Grisham's *The Pelican Brief.* Denzel Washington plays an investigative reporter looking into the murder of two Supreme Court justices. Julia Roberts is a law student whose theory about the murders has made some people very nervous. Denzel and Julia are thrown together by circumstance and a conspiracy that wants them both silenced. I've seen enough movies to know that a romantic back-story is as much a part of the action-adventure genre as popcorn and Junior Mints. So I was quite surprised when the credits started rolling, and there hadn't even been a kiss (much less the musical montage of flirting and steamy sex I was expecting) between the two leads.

Apparently I wasn't the only one disappointed. When *Newsweek* asked Julia Roberts about it in 2002, she said, "I have taken so much s—t over the years about not kissing Denzel in that film. Don't I have a pulse? Of course I wanted to kiss Denzel. It was his idea to take the damn scenes out." According to Hollywood scuttlebutt (which Mr. Washington denies), Denzel avoids kissing white women on screen for fear it will alienate many moviegoers. Especially black women, who he acknowledges are his "core audience." This fear may be reasonable since reportedly some black women were loudly disapproving of love scenes between Denzel and Mimi Rogers in test screenings of 1989's *The Mighty Quinn.* The booing was enough to convince Washington and the studio to remove the offending scenes. The good news is that Julia isn't bitter; she's described Denzel as "the best actor of this generation, hands down." (We'll be learning more about Julia and Denzel later.)

The proliferation of movies featuring interracial couples is encouraging. Sometimes the coupling is the focus of the film; sometimes it's incidental. Clearly the subject is still titillating, if not outright forbidden. In any case, I'm glad it's out there. The more we see it in movies, the more we read it in books, the more we can talk about it, the less divided we'll be.

WE DARE MAINTAIN OUR RIGHTS
(ALABAMA STATE MOTTO)

It's heartening that moviemakers have become more accepting, but Hollywood is notoriously liberal. Intermarriage is still taboo to many. Thanks to the Lovings, antimiscegenation laws were no longer enforceable after 1967, but they remained on the books in many states. For some, like North Carolina, they were even written into the state constitution. Alabama was the last state in the union to repeal its ban on interracial marriage. In order to remove this relic of its racist past, Alabama held a special election in November 2000. A full generation after the civil rights movement, 40 percent of the state voted to preserve the ban. The special amendment *was* passed, so perhaps I should be looking at this glass as 60 percent full rather than 40 percent empty. But 2 out of every 5 Alabamans is a lot of support for a law that had been unenforceable for 33 years. South Carolina passed a similar measure in 1998 with only slightly better numbers. That glass was 38 percent empty.

In 1994, Alabama was the scene of another antimiscegenation debate. Hulond Humphries had been principal of Randolph County High School in Wedowee, Alabama, for twenty-five years when he threatened to cancel the prom if mixed couples attended. After this surprising announcement, a half black/half white student was understandably confused about whom she was allowed to bring as a date. She claims that when she asked Humphries about this he called her "a mistake" that he hoped to prevent others from repeating.

Humphries's threat to cancel the prom and his purported statement to the student (which he denied) ignited a figurative and literal firestorm—there was a national frenzy of media attention, and the high school mysteriously burned down. In the end Humphries was removed as principal, and the prom went on as scheduled. I assumed this was a career-ending debacle for Humphries. I was stunned three years later when he was elected superintendent of the Randolph County School District.

The Deep South doesn't exactly have the best reputation when it comes to race relations and it's easy to write off these examples as the effects of lingering racism. But I believe the opposite—people don't fight interracial relationships because they're racist, rather racism is often motivated by an intense fear of miscegenation.

Every time that white dominance has been threatened, every time a step toward racial equality has been made, the specter of miscegenation has appeared. In fact, the term miscegenation was coined in a dishonest attempt to scare voters in the 1864 election. An anonymous pamphlet titled *Miscegenation: The Theory of the Blending of the Races, Applied to the American White Man and Negro* was printed, claiming that the ultimate goal of Abraham Lincoln and his Republican Party was the complete mongrelization of the white and black races. The pamphlet was actually written by two Democratic journalists and designed to generate fear and distrust of Lincoln. The message was clear: emancipation would lead to rampant miscegenation.

Lincoln had freed the slaves, but blacks were still denied equal rights. Black Codes after the Civil War and Jim Crow laws starting in the 1880s effectively kept blacks separate and unequal. In the Jim Crow South, blacks were segregated and denied basic rights. This was a situation blacks couldn't change because they were forced to jump through almost impossible hoops in order to vote. One voting "literacy" test required blacks to recite the Declaration of Independence and Constitution from memory in order to qualify. Whites were exempt.

In theory Jim Crow laws were about racism, but as black poet James Weldon Johnson said, "At the core of the heart of the race problem is the sex problem." Indeed, during the civil rights movement an Alabama state senator claimed that repealing Jim Crow would "open the bedroom doors of our white women to black men." Another Southerner expressed his opinion in the *Atlantic Monthly*; he wrote that desegregating schools was a mistake because it would inevitably lead to interracial dating. He colorfully suggested that thinking otherwise was about as realistic as "going over Niagara

Falls in a barrel in the expectation of stopping three-fourths of the way down."

Even liberals who supported equal rights had their concerns. President Harry S. Truman was quite progressive when it came to civil rights. He desegregated the military in 1948 and fought to make federal hiring practices more egalitarian. In a speech before the NAACP, Truman said, "We must make the federal government a friendly, vigilant defender of the rights and equalities of all Americans. And again I mean all Americans." But when a reporter asked Truman if he thought interracial marriages would become more frequent, the president replied, "I hope not, I don't believe in it. . . . Would you want your daughter to marry a Negro?" Truman said that over fifty years ago. We've come a long way as a country since then, but many are still asking that question.

Still, I am optimistic. Sure, we hear about racist crimes and the Alabama election results from 2000 are sobering, but I believe that overt racism is on the decline. Americans are more likely than ever to live in racially integrated neighborhoods and living in mixed communities leads to improved attitudes toward other races and race relations in general. Blatantly racist acts and rhetoric are far rarer than they once were. Those, like David Duke, who still wear their racism proudly on their sleeve (or white sheet), are treated as oddities rather than legitimate figures.

But despite dramatic increases in racial harmony, there has been no explosion of intermarriage. The numbers are higher than they've ever been, but are still quite low. According to the 2000 Census, about 6 percent of married couples in the United States were mixed racially or ethnically. Does that mean that 94 percent of Americans are racist? I don't think so.

More so than explicit racism, what's keeping the races separate is a desire to stay within the group. Not a hatred of other groups, but a love of one's own. Individuals can respect other groups as equal, but often choose to marry someone with a similar background. Studies have shown that on average a husband and wife's race and religion have a 90 percent correlation rate. Intelligence and personality are only around 40 percent.

I didn't need a study to tell me this—it was made clear in my upbringing. My parents are quite liberal and have no problem with interracial marriage in theory, but when it comes to their own children, they hoped we would inherit their values and marry someone like them. And they haven't been shy about telling us that. We'll take a closer look at various social and biological influences on our mate choices in Chapter Six (when we discuss blue geese and lemon-scented rats).

SOME OF MY BEST FRIENDS ARE WHITE PEOPLE . . . (BUT I WOULDN'T WANT MY DAUGHTER TO MARRY ONE)

A lot of our discussion here has been about the efforts of the white majority to prevent mixing. And studies show that whites are still the least likely group to accept the idea of intermarriage. But often resistance comes from the other side as well. Many minority group members feel a strong sense of group identity and a desire to preserve their group. This is especially true when that group has been histori- cally oppressed by the white majority. In some cases dating or mar- rying a white person is seen as a betrayal. This mentality is reflected in the Denzel Washington and Spike Lee stories told earlier.

The black community is one of the most open to the idea of intermarriage. One survey found that 86 percent of blacks thought their families would be willing to accept a white spouse. But there is still resentment, especially when white women date black men. This type of pairing is three times more common than a white man and a black woman, leaving some black women worried that white women are scooping up the most eligible black bachelors. The fear—where does that leave the group?

I grew up around a similar fear. My parents are Ashkenazi Jews who grew up in the shadow of the Holocaust. They have a very strong sense of group identity, especially because the group was so recently faced with destruction. Continuing the line and preserving the group is extremely important to them. Many people in our community fret about intermarriage statistics. It's often repeated that 53 percent of Jews in America marry non-Jews. I doubt this statistic is accurate since it never seems to change, but the number itself is irrelevant. It's repeated as a warning, and it clearly reflects the fears of the Jewish community: Did the Jews survive the Holocaust only to be eliminated through assimilation?

This conflict is famously depicted in *Fiddler on the Roof*. Tevye is a poor Jewish milkman in a small 1916 Russian village. He believes that our lives are inherently unstable; like fiddlers we are all trying to scrape out some joyful music while precariously perched on the roof of life. But the anchor that keeps us from breaking our necks is Tradition. He fears that his daughters' choice of husbands will cause Tradition to fall by the wayside.

Tevye is something of an Everyparent. His struggle has been played out countless times and continues today. The world is changing and intermarriage is no longer as rare as it once was. It's a reality that will become more and more common. I believe (and will discuss further in Chapter Seven) that we are on the cusp of a mixed marriage explosion. But the old guard is not eager to let the intermarriage phenomenon reach critical mass. The Tevyes will fight it tooth and nail.

Not because they are bad people—they're not. They are loving parents who earnestly believe they are steering their children toward the right path. But they are wrong because they are unaware of the significant advantages their mixed grandchildren will possess. Advantages that will unfold over the next few chapters. The Tevyes of this world are acting out of love, out of a sincere desire to do what's best for their family. But who doesn't want grandkids who are healthier, stronger, smarter, and more attractive? (Never mind that all grandparents think their grandchildren are those things anyway.)

Of course, this book is not an instruction manual. It's not meant to help you choose a mate or find the ideal parent for your children. Nor is it meant to judge those who choose to marry within their group. I think the Tevyes are wrong, but they are entitled to their opinion. Those who are outright racist are just plain wrong, and I have no sympathy or time for them. But the Tevyes have concerns that are real—intermarriage can lead to a dilution of culture and a loss of tradition, though it doesn't have to. I think they are fighting a losing battle, but it's a battle they have every right to fight.

So why did I write this book if not to attack the Tevyes (my parents included)? The main purpose of this book is to encourage open discourse. Interracial marriage is still taboo to many people. When it is discussed there is too often a strong focus on the negative aspects: the difficult reconciliation of two cultures, overcoming serious prejudice, and so on. I am excited to be able to point out that there are considerable benefits as well. Your choice of spouse is possibly the most important decision you will ever make. There are a myriad of variables that will go into making this choice. There is no shortage of advice and opinions in this area, but sometimes we must doubt the conventional wisdom. Sometimes we must question what our parents taught us in order to pursue our own happiness.

There is also a real sense of fear in this country. Fear of discussing race or even acknowledging its existence. Ironically, multiracial and multicultural trends are on the rise in our society. We may be heading toward a true melting pot for the first time in our history. This is a very exciting time—full of new ideas, situations, and conflicts. Yet we are more afraid to discuss these issues than we have ever been.

I am not afraid. I believe that open discussion is the only way we can truly embrace our differences and live together. Sweeping race under the rug is not a solution; it's a stalling tactic born out of fear. We can't avoid these issues forever so we might as well address them now. I'll discuss this further in Chapter Four, when I open the Pandora's Box of scary race questions.

The fact that interracial people have significant genetic advantages is fascinating to me, and I hope it will be to you. Along the way we discuss many recent and exciting scientific discoveries about human behavior. A book about science doesn't have to be boring. When I mention genetics to people, their eyes usually glaze over as they imagine complicated math, complicated chemistry, or a Moravian monk named Mendel. The discussion of female orgasms and smelly T-shirts in the next chapter will reveal that biology can be stranger, and more fun, than fiction.

With all that said, let's dive in. We'll start with a trait you've probably never thought about, but one that can predict everything from how smart you are to how good you are in bed.

2

The Leaning Tower

WHY SYMMETRY IS SEXY

THE LEANING TOWER of Pisa was not supposed to lean. Like most towers, it was supposed to be straight. The architect designed it to be straight and the builders constructed it to be straight, but it is not straight. In fact, it never will be. Not only is the tower leaning 5½ degrees to the side, but it is also slightly bent in the middle. It's actually banana shaped! The tower had already started to lean during construction so one side was made slightly taller to compensate. Despite the perfection of the plans and the best intentions of the builders, the Leaning Tower is fundamentally flawed. In this case, the flaw is a huge attraction and the main source of the tower's fame, but people are rarely this forgiving of imperfection.

Like the Leaning Tower, our bodies are constructed according to a blueprint. And like the Leaning Tower, our bodies are designed to be perfect. Not perfectly straight (our bodies are full of intentional curves and angles), but perfectly symmetrical. When I talk about symmetry (and I'm going to talk about it a lot), I'm referring to a specific type: *bilateral symmetry*. If an object has bilateral symmetry, it means that it is composed of two halves that are mirror images of each other. Most chairs fit this description. If you face a chair and cut it in half from top to bottom you'll be left with two nonidentical halves, that is, corresponding mirror images. Thus you would only have to see one of the halves to know what the whole chair looks like. Not all chairs have bilateral symmetry however. An armchair may have a lever on one side to control reclining. Since the lever is only on the right side, with no counterpart on the left side, the chair does not have perfect symmetry.

It is easy to see that our bodies are designed to have bilateral symmetry. We have two arms, two legs, two eyes, and so on: one on each side of the body. You can imagine that, if you cut a person in half, starting at the top of his head, cutting down between the eyes, through the chin, down the chest, through the navel (I'll stop there), you would end up with two corresponding halves. Each half would have one eye, one arm, one leg, half a nose, and so on. Although this is getting a little unpleasant, you can imagine if you saw one of these halves you'd know exactly what the whole person looked like.

We can see how our bodies have bilateral symmetry, but we also know it from genetics. You don't have separate sets of genes for your right and left hands. They're both constructed from the same set of genes. The same internal blueprint is used for both, so they should be identical.

We are designed to be perfectly symmetrical, but (remember Pisa!) design is not the whole story. Implementation is just as important. Just as with the Leaning Tower (although hopefully not as much) our bodies often deviate from the plan during development. Most of us don't have three arms; clearly we have pretty good symmetry, but our two halves are not identical. We all have minor variations from one side to the other. Many people are aware of these slight differences and may even have a "good side" they prefer to show in photographs. Barbra Streisand, for example, insists on being photographed from the left side; she considers the left side of her face more feminine. The disparities between sides are subtle: one study found differences in left and right index fingers to be around 2 to 4 percent. These deviations exist to a greater or lesser degree, depending on the individual.

THE CONSTRUCTION CREW

So what? Should we just feel bad that we aren't perfect or can we actually learn something from symmetry? We can learn a surprising amount. To see why, let's go back to our tower analogy. The blueprint outlines a straight tower, and the construction crew follows that blueprint to create the actual building. Let's say there are two towers: The Leaning Tower of Pisa and the Straight Tower of Pisa. Both blueprints described straight towers, but they didn't both turn out that way. Which construction crew do you think did a better job? Obviously the crew that worked on the Leaning Tower had some trouble following the plan. It seems reasonable to say that the Straight Tower crew is superior.

This is not necessarily the case. It's possible that the Leaning Tower builders faced a strong wind or other outside influence that caused their tower to lean. Maybe the Straight Tower didn't face any difficult variables. But we have no way of knowing one way or the other. We only see the finished towers; we don't know what actually happened during the construction process. Since we only have the final product to go on, we have to assume that they faced similar conditions, and the Straight Tower team just did a better job.

By now you may be wondering if you accidentally picked up a book on Italian architecture, so let's get back to people. We also have a blueprint, a master plan, encoded in our DNA. As I mentioned earlier, this blueprint outlines a body of perfect symmetry. And people also have a construction crew for their blueprint. Our genes are responsible for building and maintaining our bodies.

Let's say we examine two people: Symmetrical Sam and Lopsided Lou. As you can guess from their names, Sam is more symmetrical than Lou. What does that tell us about Sam? Both Sam's and Lou's bodies were supposed to be 100 percent symmetrical. Sam's genes have done a better job of developing his body according to that plan. Unfortunately, Lou's genes have deviated from the plan more, and as a result, his body is more lopsided or asymmetrical. Lou had some problems that made it difficult for him to grow perfectly.

Who has a better construction crew? Who has a better set of genes? Once again, we don't know what happened in the womb during their actual development. The only concrete information we have is how they turned out. With that information, we can only guess that Sam has a better set of genes that was able to follow the DNA plan more effectively.

This is an assumption. Since we have no evidence to the contrary, we assume that Sam has a better set of genes. As I mentioned with the Leaning Tower, we cannot know this for sure. Just as the Leaning Tower could have faced a strong wind or poor soil, our friend Lou could have been a victim of circumstance. Maybe he has great genes, but he ran into some adverse conditions during development. For example, children with Fetal Alcohol Syndrome are significantly less symmetrical than normal children. This is not surprising. If a pregnant woman is drinking, and her fetus has to deal with a toxin like alcohol, the baby is probably going to have problems. Development is an extremely complicated and delicate process and can easily be thrown off by alcohol, drugs, malnutrition, or some other traumatic influence.

The bottom line is that we just don't know. We have no way of knowing if Lou's mother drank and Sam's mother didn't. If we have to guess who is genetically superior, it only makes sense to choose Sam. In some ways it is like a sporting event. If Sam and Lou both run the 100-meter dash and Sam wins, it would be logical to say that he is the faster runner. This is an assumption. Maybe Lou is sick, or maybe he pulled a muscle recently. Maybe he was bribed to lose the race. There are a lot of reasons why Lou might have lost, but all we have to go on is the race itself. Since Sam won the race, we assume he is the better runner. Since Sam is more symmetrical, we assume that he has the better genes.

OK, we have determined that we can make some conclusions about the quality of a person's genes from their level of symmetry. Now let's take it a step farther. Let's go back to our towers briefly. All we know is that one of them is leaning, but based on this visual cue, we reasoned that the Leaning Tower probably had an inferior construction crew. What other determinations can we make about the towers? Which tower is more solidly constructed? Which was built with better materials? Which will last longer? The better construction crew probably built a superior tower in many respects. If we had to pick the better quality tower, the Straight Tower would be the safe bet.

We can apply this same logic to Sam and Lou. We know that Sam is much more symmetrical. We already decided he has a better set of genes than Lou. Who has a better constructed body? Who is stronger? Who is healthier overall? If we had to choose, Sam would be the clear winner. A highly symmetrical person has a good set of genes. Because of those good genes he probably has a better, healthier body than a less symmetrical person.

You may be thinking, "Hold on a minute. We seem to be making a lot of assumptions here. It is a pretty big leap from symmetry to overall health. I'm not convinced." You are absolutely right to be skeptical. So far this is all just speculation. I haven't proved anything scientifically. Science is not about making wild guesses. Science is about making wild guesses and then testing them. Luckily for us, researchers have examined symmetry extensively, and many of their findings address these issues. Let's see if their research can shed some light on our theory. Bottom line: Sam seems to be healthier than Lou as far as development goes. Is he healthier in other regards?

Yes he is.

SYMMETRY AND HEALTH

Recent studies on a wide range of species have shown that symmetrical animals are significantly healthier. In a majority of the species examined, symmetry was associated with longer life, greater health, and increased fertility. Asymmetrical individuals even grow more slowly and less efficiently than their symmetrical counterparts. This effect is seen in everything from insects to humans. A study of the forest tent caterpillar moth found that more symmetrical caterpillars live longer. More symmetrical flowers produce more nectar and attract more bees. And when it comes to symmetry, we have a lot in common with caterpillars and flowers. Men living in rural Belize were measured for symmetry and asked about their life histories. The more symmetrical men had fewer serious illnesses than their less-symmetrical peers.

Among thoroughbred horses, more symmetrical individuals have been found to run faster. And you probably won't be surprised to hear that the same is true of humans. People with greater symmetry scored higher in athletic ability and had faster running times for 800- and 1,500-meter races. What's surprising is that these symmetry measurements were taken of the runners' nostrils and ears. It's amazing that a slight difference in the size of your ear lobes can spell disaster for your running career—but it follows from our Leaning Tower logic. One flaw in appearance indicates the whole structure is inferior. A construction crew that makes one mistake is likely to make others. A lack of obvious flaws means that the entire building is probably high quality. The same is true here. A person with a great set of genes will develop very symmetrical ears and will also grow a superior, athletic body.

It turns out that symmetrical men don't just have fast feet. Their sperm actually swim faster, and they have more of the little guys. This correlation between symmetry and fertility is also found in women. Women with more symmetrical breasts are more fertile. It has also been proposed that women with less symmetrical breasts have a greater risk of developing breast cancer. (Note to men: women will not accept "I was just evaluating your fertility," or "I was just evaluating your risk of breast cancer," as an excuse for staring at their breasts.)

Symmetrical people are bigger, taller, and more muscular, but symmetry doesn't only affect physical traits. Less symmetrical men are more likely to suffer from depression, although this effect was not seen in women. One study even found that symmetrical men and women score higher on intelligence tests. Finally, more lopsided individuals are more likely to respond aggressively when provoked. Maybe they're just grumpy because they know that symmetrical people are so much better off.

The philosopher Thomas Hobbes described life in the natural world as, "solitary, poor, nasty, brutish, and short." Some of the advantages of symmetry that we just discussed would appear to make life a little less well . . . nasty, brutish, and short. Resistance to disease obviously extends an animal's life, and overall athletic ability will help an animal find his dinner and avoid being someone else's. One "brutish" danger that animals face repeatedly is parasitism. Parasites sap valuable resources and can lead animals down the path of sickness and death. Do more symmetrical animals have any advantages when it comes to resisting parasites?

Yes again.

Scientists have found that individuals with greater symmetry are less likely to have parasites. At first blush this seems to answer our question, but this is a chicken-and-egg problem. It could be that more symmetrical animals are better able to resist parasites. However, it is also possible that all individuals are equally likely to be infected, but that parasites weaken animals and make them more lopsided in the process. Studies in this area indicate it's probably a combination of both. When barn swallows are infected with mites, they do become less symmetrical. However, at least in some cases, more symmetry means more resistance to parasites. This is seen in house flies where more symmetrical flies are more likely to fight off a fungal infection.

The bottom line is clear: Symmetry is good. Symmetry is associated with all sorts of great qualities, including disease resistance, fertility, athleticism, and overall health. With this knowledge, who do you think a woman would rather marry, Symmetrical Sam or Lopsided Lou? No offense to Lou, but with all of this evidence, Sam sounds like a catch. Having a baby is a big investment, and animals, including humans, have evolved to increase the chances of baby survival any way they can.

There are several possible advantages to having a symmetrical mate. In many species, like humans, both parents are involved in caring for their child. Being a single parent is difficult now, but in the old days it could spell doom for the remaining parent and child. Symmetrical individuals have significant survival advantages. By picking a symmetrical mate, you can decrease the chance of your mate dying, leaving you to fend for yourself and your baby alone.

However, in many animals only one parent raises the offspring. It is quite common for the male to be nothing more than a sperm donor. He performs his duty and is quickly off in search of new opportunities. This is the case for peacocks. A male peacock dazzles a female with his resplendent tail, mates with her, and then he and his tail hit the road looking for more "chicks."

This is quite a contrast to the sex lives of giant water bugs. Here, the father supplies all of the parental care. During mating, the female glues the eggs to the male's back, and he is left with the onerous task of caring for them. With the glue still wet, she flies off in search of other males.

In other cases there is no parental care at all. This is true of most species of fish. The female lays her eggs, the male fertilizes them, and then they both swim off. No diapers, no carpool, no college tuition. However, even when parental care is not an issue, a symmetrical mate is still desirable. If nothing else, a parent always contributes at least one thing: genes. Even though fish don't seem very nurturing, they still want their babies to survive and thrive. The best way to do that is to start them off with a good set of genes.

Sexual reproduction is a joint venture. Each parent contributes half of their genes, and the child is a composite of these genes. In effect, it's the creation of a new tower. Half of the builders for the Baby Tower will come from your construction crew, and the other half will come from the crew of your mate. Since you want the new tower to be as solidly built as possible, you want to assemble the best possible construction crew. Would you take half of the new crew from the Leaning Tower or the Straight Tower? Once again the answer is obvious.

You want your child to succeed so he needs high-quality genes.

A person's symmetry indicates their good genes. They are the best prospect for a joint venture. So, if symmetry really does indicate all of the great attributes we talked about, wouldn't animals have evolved to seek out symmetrical mates?

They have.

Hey Baby, What's Your Symmetry?

Symmetry and its effect on mating have been studied extensively. Researchers recently performed a meta-analysis of the many experiments on this topic. They analyzed the data of 146 samples from 65 studies of 42 species ranging from insects to mammals, and found a significant relationship between symmetry and attractiveness to the opposite sex. Animals seek out symmetrical mates, and humans are no exception. In experiments, men and women consistently rate symmetrical faces as more attractive. In one study, people were shown two photographs of the same person. One was a regular photo; the other was manipulated by computer software so the face was perfectly symmetrical. The differences between the two photographs were so subtle that the judges claimed to be unable to distinguish between them. But when they had to decide which was more attractive, they chose the symmetrical one far more often.

The really interesting trends, however, take place outside the lab. Symmetrical men lose their virginity 3–4 years earlier and have 2–3 times as many sexual partners as less symmetrical men. Symmetrical women also have more partners than lopsided women. It's good to be symmetrical! Symmetrical men also have sexual relations earlier within a relationship. But women beware. Not only are symmetrical men likely to be more experienced, they are also more likely to cheat on you. Symmetrical men have more sexual relations outside of their relationships. But before you start measuring your mate, keep in mind that this doesn't necessarily mean that symmetrical men are more willing to cheat than other men.

It may just be that because of their higher attractiveness, they have more opportunities to do so.

In a fascinating study at the University of New Mexico, researchers examined the sex lives of undergraduate heterosexual couples. This sounds like a study from the University of Melrose Place, but the findings are remarkable. The participants filled out detailed questionnaires about their sex lives and were measured to determine their symmetry. Analysis of the data revealed that women with more symmetrical partners were much more likely to orgasm during sex! The man's age, attractiveness, sexual experience, socioeconomic status, and perceived future earnings had no effect on his partner's likelihood to orgasm. Nor did the length of the relationship or the participants' feelings of love. Male symmetry, however, was a significant predictor of female orgasm.

This is certainly an exciting finding (particularly for women with symmetrical partners), but what does it mean? Why would symmetrical men be more likely to bring women to orgasm?

Remember what Theodosius Dobzhansky said, "Nothing in biology makes sense except in the light of evolution." It's unlikely that he was thinking about women's orgasms when he said this, but it's applicable in any case. Evolutionary theory does provide the answer here.

There is some evidence that female orgasm can increase the chances of conception. Despite the experiences of some girls at my high school, impregnation is not achieved easily. Each sperm is only 1/500th of an inch long, and the route through the vagina and uterus to the waiting egg can be as long as a foot. This may be one of the reasons that males have penises. When released from the tip of an inserted penis, sperm are already part of the way to their goal. But reaching the egg is still no walk (or swim) in the park. Especially because, contrary to popular belief, sperm are not very good swimmers. Of the 300 million or so sperm (roughly as many as the population of the United States) that are released in an average ejaculation, only 2,000 (roughly as many as the population of my high school) even make it to the egg. Not very good odds.

Here is where a female orgasm can make all the difference. Before orgasm the uterus has positive air pressure, but during orgasm, muscle contractions cause a sharp change to negative air pressure. This creates a vacuum, which sucks the sperm up into the uterus. This increases the chances that sperm will reach the egg and fertilize it. This is known by the lovely name of "upsuck" theory.

So let's say Fertile Fran is playing the field and sleeps with both of our friends, Sam and Lou. We've already established that women should want symmetrical men to father their children. Clearly Sam is the more desirable candidate, but if Fran sleeps with both of them, there's only a 50 percent chance that he will be the father, right? Wrong. If she has an orgasm when she's with Sam, the upsuck will give him much better odds of impregnating her. Lou is less likely to give her an orgasm so his sperm are on their own.

This is a way that women can try to "choose" the father of their children. If a woman is mating with more than one man she will "want" the more symmetrical one to impregnate her. By preferentially orgasming with him, she increases the odds that fertilization will happen. Now, of course, Fran isn't thinking, "Boy, Sam sure is symmetrical. His developmental stability indicates high-quality genes. I want my offspring to have high-quality genes too. By orgasming now I can improve the odds that he will impregnate me and I will have a healthy baby. OK, here we go: three . . . two . . . one."

That's just silly. Women don't think that way, especially not in the middle of sex. At least I hope not. But we have evolved so that on some level this logic is taking place. Fran isn't consciously thinking about it, but her body takes care of it automatically. In fact, Fran and Sam may even be using a condom, in which case she isn't upsucking anything. But her body doesn't realize that, it evolved in a world without contraception—where every sex act could potentially lead to pregnancy. Her body may not understand condoms, but it has evolved some pretty neat tricks.

THE SWEET SMELL OF SYMMETRY

Sometimes it isn't even necessary to see a potential mate to decide how desirable they are. Female Japanese scorpionflies are more attracted to symmetrical males. This by itself is not very exciting. It's consistent with the pattern we have seen in many other animals, including humans. What is interesting about the scorpionfly is that the female is more attracted to a symmetrical male without ever seeing him directly. She makes her decision based purely on his smell. He releases special pheromones to attract females, and more symmetrical males do better, even sight unseen. This is pretty neat and it fits in perfectly with our theory. Remember, we are not saying that there is anything innately good about symmetry. Symmetry is only a desirable trait because it gives us information about genetic quality. Since we cannot see genes directly, we use symmetry to guess what's going on in an individual's DNA.

Symmetry is like the fuel gauge in your car. I used to drive a car with a broken fuel gauge. Let me tell you, every drive was an adventure! It's such a simple thing, we take it for granted, but when my gauge was broken I really missed it. I didn't miss it because it was exciting or fun to look at. The needle itself doesn't do anything for me. But where the needle points does matter to me because it gives me information about something that I care about. I care a lot about the amount of gas left in my tank. A few times it made the difference between a relaxing evening drive and a sweaty evening push. My gas level is important information. Information that I can't get directly because the gas tank is buried inside the car.

Similarly, we can't assess the quality of a person's DNA directly because it is buried inside their cells. So we use other cues that we can see. You can think of symmetry as a quality gauge. It's as if we all walk around with a gauge on our foreheads where everyone can see it. A symmetrical face and body are like a gauge with the needle pointing to FULL. Anyone can see that your genes have done a good job developing a stable body. A more lopsided figure means the needle is leaning more toward EMPTY. Bottom line: the gauge is not important itself, but the information that it provides is.

Multiple gauges can give you the same information. The female scorpionfly doesn't use visual information alone to decide about a potential mate. She also uses his scent. In this case there are multiple ways to get the desired information. (I wish there had been for my car.) A male scorpionfly's symmetry is one source of information, but the quality of his scent is just as valid.

Sniffing out a good mate may sound pretty outlandish, but people do it too. We can see this phenomenon in another innovative study from the University of New Mexico. The researchers that brought us the orgasm study we talked about earlier also did this experiment, so you know it's going to be good.

The goal of this study was to see if women can do what scorpionfly females do: pick out symmetrical, high-quality males based on smell alone. The experimenters recruited forty-two men to contribute their scents. They measured each man for symmetry and then gave him a new, white T-shirt to wear. Each participant slept in his new T-shirt for two consecutive nights. To avoid contaminating the shirts with other scents, the men followed strict rules. Use of deodorants or colognes was not permitted. In addition, the men refrained from smoking, didn't eat any stinky foods like garlic or onions, and slept alone. After the second night each man placed his shirt in a plastic bag and returned it to the experimenters.

The second phase of the study began right after the shirts were returned so that they were still "fresh." Fifty-two lucky women evaluated the smell of each shirt. Each woman opened each bag, took a nice big sniff, and then rated that shirt on pleasantness, sexiness, and intensity. The women also filled out a questionnaire with basic background information, including whether they were on the birth control pill and the first day of their last menstrual period.

You can probably guess the results of the study. You may be getting tired of all of these studies where the symmetrical guys win hands down. But this one has an interesting twist. The shirts worn by the more symmetrical men were rated higher, but only by some of the women.

The women who greatly preferred the symmetrical shirts were around the time of their ovulation. Women who were at less fertile points in their cycle and women who were on the pill liked all the shirts about the same. The higher the likelihood the woman was ovulating, the stronger the preference for the symmetrical scents.

This is an interesting result. Not only can women sniff out a more symmetrical man, but also they only do it when they are most fertile. This isn't entirely expected, but it's easily explained by our theory. For the past few pages we've been discussing the fact that, evolutionarily, a woman should prefer a symmetrical man. Why? Because if a symmetrical man is the father, the child will be of better genetic stock. Evolutionarily, women want to be impregnated by symmetrical men. These preferences only matter when the woman can actually get pregnant. That's exactly the trend we see in this study. When a woman is fertile, she's very sensitive to the symmetry of a man. This is the time when she needs to be sensitive so she can pick out the best man to have a baby with. When a woman is not at a fertile stage, it doesn't matter as much. Since she is unlikely to conceive, preference for symmetry is not important, and it disappears.

The birth control pill suppresses ovulation so a woman has a greatly reduced chance of getting a bun in her oven. Her body knows there's no chance for conception, with any man, and the symmetry preference disappears.

Why not have the symmetry preference all the time? The sensitivity to symmetry may be difficult or require a lot of energy to maintain. Since we evolved in a world where food was scarce, our bodies try to be as energy efficient as possible. If the preference is costly in some way, it makes sense to only bring it out when it is needed.

It's a lot like shopping for a car. When you are in the market for a new car you pay a lot more attention to the cars around you. You examine them more closely, checking out ones that are better suited to you. You have a strong preference for some cars. When you are not looking for a car, that greater preference disappears. You don't pay as much attention to the cars on the street because they don't have as much relevance for you. You don't bother differentiating between Accords and Camrys. A Lamborghini might still catch your eye, but, in general, cars just blend into the background.

When a woman is fertile or "in the market" for a man, she has a strong preference for the better, higher quality models (the BMW, not the Yugo). When she isn't fertile, she doesn't waste time and energy focusing on something that doesn't affect her. All the men just sort of blend together, no one really stands out. It's another example of our "smart" bodies at work. Of course a woman isn't consciously thinking about these things. Many women are not even aware when they're ovulating, but their bodies "know" and act accordingly.

SYMMETRICAL MEN AND THE WOMEN WHO LOVE THEM

I've told you about women having orgasms with symmetrical men and women sniffing out symmetrical men. What about men sniffing out symmetrical women? The fact is that most of the research in this area, and most of our discussion here, is about men's symmetry and its effect on women. Why the focus on men? It may seem like there is some sexist conspiracy in the scientific community. Perhaps scientists all think of women as baby-making machines perpetually searching for a mate. That's ridiculous. The truth is that both women and men are baby-making machines perpetually searching for a mate!

One of the strongest desires of both sexes is to reproduce. Evolution has ensured that. We are all descended from people who wanted to and did have children. Remember our ancestors, the sex-loving cave people? This craving is so strong that the few individuals who don't have children are often pitied. It is assumed they just didn't have the opportunity. People who claim not to want children are often regarded with surprise or even suspicion.

We have evolved to have as many children as possible. Both men and women are designed to maximize their reproductive potential. This is why we are so drawn to the opposite sex. This is why sex is fun. These motivations encourage us to reproduce, reproduce, reproduce. Our modern-day priorities are not exactly in line with this, and few of us have as many children as we could. But we are still influenced by those evolutionary urges. We are still drawn to the opposite sex, and we still find sex fun.

Since men and women both want to have children, why the focus on male symmetry? Although both sexes are similar in their desire for reproduction, they are different in how they seek it out. Men and women are different—no surprise. Countless books, magazine articles, and television shows have been devoted to that very subject. And this should come as even less of a surprise: when it comes to sex, women are pickier than men.

Comedian Lenny Bruce once said, "Men will fuck mud." This is quintessential Lenny Bruce: vulgar and exaggerated, but with a core of truth. The implied counterpoint is that of course a woman would never have sex with mud—women are much too selective. This is so ingrained in us that men and women are often judged by different rules. A promiscuous woman is looked down upon as a "slut." She's not being as selective as she should be. Men, on the other hand, aren't expected to be picky. Thus, a promiscuous man is admired as a "stud."

To Lenny and to most of us this is obvious, but in science nothing can be assumed and even the slut/stud divide has been put to the test. In a study of college students, men were approached by an undercover female experimenter of average attractiveness. She offered casual sex with no strings attached. Three-fourths of the men thought it was their lucky day and said yes. They were surely disappointed to discover it was just an experiment. In the second phase of the study, an average looking man made the same indecent proposal to female college students. How many women do you think took him up on it? None. This experiment is repeated every day at your local singles bar, with similar results. Our suspicions are confirmed. Women are pickier, but why?

As Dobzhansky advised us, we must look to evolution. Why have women evolved to be picky and men evolved to be easy? The simplest explanation is that women get pregnant. Both men and women want to have children, but the responsibility of actually having them rests entirely with women. This results in completely different reproductive strategies. Both sexes are trying to maximize their offspring. A woman has only one child at a time. (Twins, triplets, etc. are a rare exception.) This is a commitment of several years per child. A man's commitment, on the other hand, is sometimes only a few minutes. Is this fair? Of course not! But evolution is not concerned with fairness; each individual is looking out for him/herself. Everyone's trying to increase their own chances of survival and reproductive success, sometimes at the expense of others.

So let's return to Fertile Fran. She is choosing between Symmetrical Sam and Lopsided Lou. She has to choose because she can only have one baby at a time. Since she can only have a baby with one of them, she has to choose either Sam or Lou. Her choice takes place on multiple levels, both obvious and subtle. She is more attracted to Sam so she is more likely to mate with him. But even if she sleeps with both of them, she is more likely to be impregnated by Sam because her body will help his sperm reach the finish line.

Let's reverse the scenario. Fertile Fred runs into Symmetrical Samantha and Lopsided Louise, and they're both ready and willing to sleep with him. (Fred is a lucky guy. Probably very symmetrical since he's getting multiple offers.) If he had to choose one, it would no doubt be Samantha, but because of the way human reproduction works, he doesn't have to choose. If he really wants to maximize his reproduction (and evolution has ensured that he does), he should try to impregnate both of them.

This is a simplified view of things. The truth is that men often provide parental care and don't always jump from woman to woman. But women will always be choosier because the costs of a pregnancy are so high for them. A child is an equally huge benefit for both parents, but the costs are far from equal. The high cost of a pregnancy means that a woman should be selective. A man, however, has sperm to spare; his costs are so low that he would be foolish to turn down a chance at reproduction. It's not that symmetry is only meaningful for men. Symmetry is a hallmark of quality for both genders. However, while women actively seek out quality, men look for quantity.

An old joke sums this up perfectly:

> *Question:* What is the difference between a woman and a man?
> *Answer*: A woman wants one man to satisfy her every need. A man wants every woman to satisfy his one need.

These cost/benefit calculations may seem unfair to women, but they are an inevitable consequence of the way we reproduce. It has nothing to do with actually being female. The gender with the greater cost will always be choosy. And the gender with lower costs will always be promiscuous. Perhaps my female readers can take some consolation from the giant water bug that I mentioned earlier.

Giant water bugs' gender roles are the opposite of humans'. This is because the costs of reproduction are higher for the male. When water bugs mate, the male fertilizes some of the female's eggs. That part is pretty standard, but then things get interesting. Instead of caring for the fertilized eggs herself, she glues them to his back, and he must tend to them until they hatch. The eggs are heavy so he can barely swim, and the glue seals his wings closed so he can't fly. In effect, he's "barefoot and pregnant." As you might expect, he is very choosy about mating and doesn't take on this responsibility recklessly. The female doesn't have to take care of the fertilized eggs so her costs are low. She swims around gluing eggs onto any male she can grab a hold of. This unusual distribution of costs completely changes the dynamics of mating. Imagine how differently men would behave at singles bars if they could end up with babies glued to their backs at last call!

I don't know if it's been studied, but I'd guess a male water bug would much prefer to mate with symmetrical females. Since his reproductive potential is limited by his costs, it would make sense for him to focus on mating with high-quality, symmetrical females.

UNRESOLVED QUESTIONS

Symmetry is a powerful concept. The greater health and attractiveness of symmetrical individuals make it a very desirable trait. We've established that symmetry is a hallmark of fitness and beauty, but some questions remain to be answered. Those questions fall into two categories: questions that I'm not going to answer and questions that I'm going to answer later.

QUESTIONS I'M NOT GOING TO ANSWER

There are some interesting aspects of symmetry that I haven't gone into and that I'm not going to go into. One area of symmetry that I'm particularly interested in is how exactly we measure each other's symmetry. We can't calculate symmetry consciously, but we are aware of it on a subconscious level. How exactly do we measure it? What parts of the face and body do we take into account? I wish I could tell you how it works, but we don't know yet. I think that others, especially people in the fashion industry, will be interested in this information as well. After all, once we know how this process works we can begin to manipulate it to make ourselves seem more attractive.

We already use makeup to take advantage of evolutionary preferences. Clear, unblemished skin was historically associated with a healthy individual, free of diseases and parasites. People with flawless skin are more attractive since it's a hallmark of health. In the modern day we use makeup to give us the illusion of clear skin and thus make ourselves seem more attractive. Could we do the same with symmetry?

Symmetry is a lot harder to fake than skin tone, but there are definitely possibilities. We already do some things to improve our symmetry. Corrective orthodontia makes our teeth and smile more symmetrical. Clothing probably hides some of the subtle asymmetries of our bodies. And symmetry is factored into any cosmetic surgery.

Should I part my hair in the middle because it's more attractive than a more lopsided hairstyle? Maybe or maybe not. It is possible that hair makes no difference. If you have a scar or blemish, is it better to have it in the middle of your face so that it doesn't affect your symmetry? Hard to say. Cindy Crawford's mole is an obvious asymmetry yet it doesn't detract from her beauty. There are a lot of exciting questions here, but we won't know the answers until we understand the mechanisms that we use to calculate symmetry.

Questions I'm Going to Answer Later

But I still have a lot to tell you. This book is about interracial people, but I haven't mentioned them at all in this chapter. What does symmetry have to do with them? We'll get to that very soon.

Also, where does symmetry come from? We've established that symmetry is an indicator of health and because of that a very attractive characteristic. But since all bodies try to develop with perfect symmetry, why do some get closer than others? I touched on this a little bit. I said that symmetrical individuals have better genes that allow them to develop with more stability. This makes sense, but it's a little vague. What exactly are "better" genes? Why would a gene work differently on the two sides of the body? How do people get these "better" genes? These are important questions and the answers are coming. In the next chapter I'll go into more detail about "better" genes, how they work, and why these genes give some lucky individuals increased symmetry and the sexiness that comes along with it.

☞ 3 🖎

Building the Tower

GENETIC VARIATION DOES A BODY GOOD

IN 1722, CAPTAIN Jacob Roggeveen, a Dutch explorer, hit shore on the most isolated inhabited island on earth. Since he landed on Easter Sunday, Captain Roggeveen named his remote discovery Easter Island. At the time, the island was home to a few thousand natives and 887 giant stone heads, and Roggeveen wasn't sure how either the natives or the heads got there. Since that day, Easter Island has been associated with unsolved mysteries. The island is truly in the middle of nowhere, over two thousand miles from both South America and Tahiti. Where did the natives come from? Most archaeologists now believe they descended from Polynesians who sailed to Easter Island in double canoes. But Norwegian anthropologist Thor Heyerdahl was so certain that the original settlers were from South America that he sailed his balsa raft, *Kon-Tiki*, over four thousand miles from Peru just to prove it could be done.

Even more puzzling were these giant stone heads, called *moai*. Who built them and why? And how? The largest fully erect *moai* is 32.63 feet tall and weighs approximately 82 tons. How did primitive people build and transport such massive statues? There is no shortage of theories. Modern anthropologists have tried to replicate this feat using only ancient technology, with mixed results. Erich Von Daniken proposed that the statues, which stand on pedestals looking out to sea, were the work of aliens. But one thing is clear, someone built these *moai*. Statues don't just appear. And neither do towers.

Towers don't just appear; they must be built. Just like the *moai*, every tower has a construction crew. In the last chapter we saw how important the crew is to the quality of the finished product. So let's take a closer look at this construction crew. For the sake of preparedness, let's say that our crew is composed of two workers of each type: two carpenters, two masons, two stonecutters, and so on. Which will result in a better tower: two carpenters that turn out exactly the same product or two carpenters that have different skills and training? Building something as complicated as a tower involves many different situations and conditions. There's a clear advantage to having two different carpenters because their diversity of skills allows your crew to handle a variety of challenges.

I've already mentioned that our bodies also have a construction crew: our genes. And you won't be surprised to hear that we have two of each type. You have two copies of every gene, one inherited from your father and one from your mother. These genes produce the proteins that are both the raw materials and construction crew of our bodies. The way your parents' genes combine and interact, with each other and with your environment, determines the ultimate shape and condition of your body.

In *Cat's Cradle*, Kurt Vonnegut satirically explores the end of the world and an enigmatic religion named Bokononism. The religion's bible, *The Books of Bokonon*, opens with, "All of the true things I am about to tell you are shameless lies." This is typical of the perverse logic of Bokonon. At one point the narrator writes, "*Busy, busy, busy*, is what we Bokononists whisper whenever we think of how complicated and unpredictable the machinery of life really is." I think he was referring to philosophical issues like fate and the interconnectedness of all things, but he might as well have been thinking of the literal machinery of life.

Genetics is simple in theory. We have two copies of every gene. For the most part, each gene produces a protein that serves some function in developing and maintaining our bodies. Simple, right? But our bodies operate on a massive scale. Every day, just to keep your body running, you produce billions of fresh proteins. Billions! These proteins come in a staggering forty thousand varieties and have to be produced in the correct quantity, at the proper time, and delivered to the appropriate location. Whether you're fighting off an infection, digesting some food, yelling at someone in traffic, or just daydreaming, your genes are busy, busy, busy.

As I've mentioned, our busy genes come in pairs. With two copies of each gene, there are only two possible scenarios: the two copies are identical (you are homozygous for that gene), or the two copies are different (you are heterozygous for that gene). If your two copies are different, they still serve the same role, but the proteins they code for are not the same. They are often similar, but not identical.

I'll ask the opening question again with our new terminology. Would you rather be homozygous for a gene or heterozygous? Don't be confused by what we talked about in Chapter Two. You definitely want your body and face to be as symmetrical as possible, but when it comes to your genes, asymmetry is good. Your gene crew is building and operating a very sophisticated body that may have to function in the blistering Sahara or the frigid Arctic. Having two different proteins at work can give you some advantages under varying conditions. We'll talk about how these advantages might work in a little bit.

BUT ISN'T ONE OF THE GENES DOMINANT?

Let's take a step back. If you're like me, in seventh grade health class, in addition to watching a filmstrip called *Becoming a Man*, you learned that we have two copies of every gene, and these copies can be dominant or recessive. The dominant genes overpower the recessive genes, so if you have one of each, only the dominant one will be expressed. While this isn't a complete load of hooey, it's not relevant the vast majority of the time. So why did Mrs. Griffin teach it to me? Well for one thing, it's simple. And for another, dominant and recessive genes *are* relevant in an area that people care a lot about: diseases.

Lazy Carpenters and Rotten Wood

Genetic disorders are caused by defective genes. These aren't diseases that you can catch like viruses or bacteria; you're born with them. Your parents have defective genes (no offense—we all do), and there's a chance that you will inherit some of them, which may lead to an actual disorder. Some genetic disorders are recessive, which means that you'll only have the disease if both of your copies are defective. If you have even one good copy, you'll be fine. If you have two good copies, even better. But if you have two bad copies, you express the disease. A recessive disease gene is like a lazy carpenter. If you have one lazy carpenter in your crew, your tower will probably be fine because you have another, industrious, carpenter to pick up the slack. The good carpenter might have to work overtime, but the job will get done. If you have two lazy carpenters, however, the work won't get done, and your tower will have some serious shortcomings to deal with in the woodworking department—shortcomings that can be disastrous.

As in our carpenter example, recessive diseases often affect genes in charge of *doing* things. These genes produce enzymatic proteins. Enzymes are the workers of the body; they make things happen. If your body is breaking down food, growing new cells, or transporting nutrients throughout the body, enzymes are doing the bulk of the work. If one copy of an enzymatic gene is defective, often the other can still shoulder the burden. If both are defective, the work goes undone, and (depending on that particular enzyme's job) serious consequences can result.

My father has a recessive genetic disorder called Gaucher disease. Our bodies have a gene that produces an enzymatic protein called glucocerebrosidase. This is a fancy name for what is basically a garbage man protein. This enzyme has one job—it breaks down a certain type of fat molecule. That's it: no more, no less. Inside every one of our bodies these little guys are hard at work—breaking down these fats. Well, inside every one of our bodies except my dad's.

He inherited two defective copies of the glucocerebrosidase gene from his parents—he's homozygous for the defective version. Without any functional copies, he can't produce any glucocerebrosidase proteins that actually work. This means that these fat molecules build up in his body and can cause problems with his liver, spleen, and bone marrow. He basically has two lazy garbage men. They don't take out the trash, it builds up, and problems ensue. Happily for my dad, this genetic disorder is pretty mild. His blood doesn't clot very well (his bone marrow has trouble making platelets because it's clogged with fat), but for the most part he's fine.

I'm what's called a carrier for the disease. I inherited one bad copy of the gene from my dad, since he only has bad copies to give. But luckily my mom was nice enough to give me a good copy of the gene. With one lazy garbage man and one hardworking one, I do just fine. My fat molecules get broken down and carted away properly. I'm called a carrier because I do carry a copy of the disease gene, but I don't have the disease because I also have one good copy—I'm heterozygous.

My heterozygosity protects me from suffering the ill effects of Gaucher disease, but that's not always the case. Some genetic disorders show symptoms even if only one copy of the defective gene is present. These are called dominant disorders. Often these disorders involve the structural proteins that compose the framework of our cells. We can get by with one lazy carpenter, but dealing with flawed structural materials is another matter. If half of the wood that our carpenters are using is rotten, the tower is going to run into trouble, even if the other half of the wood is fine. In cases like this, having one good copy of the gene can't protect us from the negative effects.

I'm happy to report that my father does not have Huntington's disease. People who do suffer from Huntington's have a defective copy of the gene that produces Huntingtin. The function of the normal Huntingtin protein is not known, but the abnormal Huntingtin produced in individuals with a defective copy of the gene starts killing brain cells when the person is thirty to fifty years old. This leads to dementia and wild involuntary movements. The disease is also called Huntington's Chorea because of the uncontrollable writhing and twisting motions associated with it. (Chorea is the Greek word for dance—same root as choreography.) But don't be fooled by its lighthearted name—you could call it Huntington's lambada, it wouldn't change the fact that even one defective copy of this gene is a death sentence. With a dominant disorder like Huntington's, heterozygosity doesn't offer any protection.

In some cases it's possible that being heterozygous for a genetic disorder may not just be neutral, but may actually provide some benefits. This may help explain why certain defective genes are so common. Even if two copies of a gene are harmful, if one copy is helpful, evolution may still favor the gene. This is called heterosis or heterozygote advantage.

The classic example of heterosis is sickle cell anemia. Red blood cells with the sickle cell gene have a slightly different version of the hemoglobin protein. When faced with low oxygen concentration, they have a tendency to lose their round shape and become sickled. This can cause a disastrous cycle when sickled red blood cells get stuck in small capillaries. This causes a blood cell traffic jam, depriving the region of oxygen and in turn causing other red blood cells to sickle.

If you have two copies of the sickle cell gene, then all of your red blood cells have the tendency to sickle. This can cause serious health problems, including significant damage to the spleen and other internal organs or stroke.

But it turns out that people with one copy of the sickle cell gene have an advantage. They are resistant to malaria. The malaria parasite spends part of its life cycle inside red blood cells, but the parasite cannot hijack red blood cells with the sickle cell trait. Since carriers also have a normal version of the gene, many of their red blood cells are normal. They experience few of the health problems associated with sickle cell anemia, but enjoy malaria resistance.

It's easy to imagine that where malaria is present, sickle cell carriers have a big advantage. They stay healthy and have more children, thus passing the sickle cell gene on to the next generation. Unfortunately, sometimes a child will end up with two copies of the gene and suffer from the disease. But the strong advantage of the heterozygous carriers will keep the gene in the population.

It's not surprising that the sickle cell anemia gene is common in people who originated in areas where malaria is a serious health risk, like the Mediterranean, India, and Africa. The disorder is especially common in West Africa. In fact, sickle cell anemia is the most common genetic disorder affecting African Americans, most of whom are of West African descent. Approximately one in every twelve African Americans carries the sickle cell gene. Of course, Americans of African descent are no longer at risk of being infected with malaria. The protection offered by the sickle cell gene is moot, but they are still saddled with the genes of their ancestors and the related risks.

A similar situation has been proposed concerning the most common genetic disorder affecting Europeans—cystic fibrosis. If you're of European descent, there's a 1:25 chance that you carry the cystic fibrosis gene. This gene codes for a protein that's involved in maintaining water levels in the body. Two defective copies of the gene is bad news. Sufferers of cystic fibrosis produce overly thick mucus in their lungs and intestinal track and are extremely prone to respiratory infections. There are ways to manage the disease that involve loosening and removing the mucus with the aid of inhaled medication and vigorous chest pounding, but life expectancy is still only thirty-two years. Individuals afflicted with cystic fibrosis are *too* good at keeping fluids in their bodies.

But in the face of dehydration, this can be a valuable trait. Historically, diseases such as cholera often ended with death by dehydration caused by severe diarrhea. It's pretty disgusting, but the body loses fluids so rapidly that it is almost impossible to replenish them. Carriers of the cystic fibrosis gene may have been able to hold onto their water more effectively and ride out the disease. This type of advantage may have kept the cystic fibrosis gene in the population.

A QUICK LOOK AT MY BRAIN

Recently, Gregory Cochran, an independent scientist, and some researchers from the University of Utah proposed that heterosis may also be involved in some of the genetic disorders affecting Ashkenazi Jews. According to their theory, Jews in the Middle Ages were prohibited from owning land and were forced into industries like banking and trade. These are professions that place a high premium on intelligence. The ancestors of these Jews score 12 to 15 points above average on IQ tests and coincidentally (or maybe not) suffer from a family of similar genetic disorders including Tay-Sachs, Gaucher, and Niemann-Pick disease. We've already discussed Gaucher, and it turns out that Tay-Sachs and Niemann-Pick also involve the buildup of certain types of fat molecules. These fats are an important component of nerve cells, specifically of the insulating sheaths on the outside of neurons. These sheaths are involved in sending electronic signals, which is how your nerve cells communicate. Cochran and his colleagues speculate that carriers for these disorders have more developed knobs on their neurons, which allow their brains to be more tightly linked and thus more intelligent.

These scientists are basically saying that dumb Jews are like slow cheetahs—they didn't have as many kids and didn't pass their genes along. Smarter Jews, who also may have been carriers for these diseases, did pass their genes along. And those are the genes that modern-day Jews have inherited. If you were paying attention earlier, you realize that this is not academic discussion for me; Cochran is talking about my brain! He claims that being a carrier for Gaucher disease has given me an additional 5 IQ points.

This is interesting stuff (especially for me), and Cochran did find records showing that more financially successful Jews had more children historically. There is also some anecdotal evidence from a Gaucher clinic in Israel that more of its patients are scientists, engineers, and lawyers than the regular population. Still, I'm not signing up for *Jeopardy!* just yet. It's an interesting theory, but it's very easy to speculate about evolution after the fact, and far more difficult to prove these speculations. In fact, other scientists have previously suggested a different theory of heterosis for Tay-Sachs disease. There is some evidence that carrying the Tay-Sachs gene may provide some protection against tuberculosis. This would have been extremely valuable to Jews forced to live in dense urban ghettos where tuberculosis was commonplace.

Examples are difficult to prove, but recessive genetic disorders do provide some benefit under certain conditions. Suffering from the disorder can be devastating, but those merely carrying one copy of the gene can enjoy significant benefits without incurring the costs of the disease.

This was a long digression to explain what dominant and recessive mean. The truth is, however, that most of the time they're irrelevant. Genetic disorders are extreme cases involving defective genes, but most of the time we're dealing with different, but functional, versions of genes. If you have two competent carpenters with different skills, they are both working side by side. Neither is masking the other. Similarly, most of the time heterozygosity means two different versions of the same protein both working simultaneously. One is not dominant over the other.

So What?

If both versions of the gene work, what's the big deal with heterozygosity? The key is that both versions of the gene, and the resulting protein, work, but not necessarily in exactly the same way. For example, every enzyme has an environment in which it works best. The enzyme is at peak performance at a specific temperature and pH. If the environment deviates from the ideal, the enzyme functions less efficiently.

If an enzyme's ideal temperature is 100 degrees, it will function less efficiently at 101 degrees. If you are homozygous for that gene, you only have one version of the enzyme. If the temperature goes up (or down), your enzyme isn't working as well as it could. That can be a problem. Especially because the environment can be different in different parts of your body. What if this enzyme is involved in growth and development and the left side of the womb is a little warmer than the right? The enzymes on the right side of the body are working at maximum capacity to stimulate growth. But on the left side the increased temperature means the enzymes are not keeping pace. Less growth on the left side means an asymmetrical body. And you know that's not good news.

A different version of the same gene may produce a protein with a different ideal environment. Let's say this version is optimized for 102 degrees. If you're heterozygous and you have both versions, you have more flexibility as far as temperature goes. As the temperature rises, version 1 starts operating less efficiently, but at the same time, version 2 kicks in and balances things out. Being heterozygous for this gene means you have a wider temperature range at which your enzymes can operate efficiently. Minor temperature fluctuations aren't disruptive for you like they are for your homozygous friend. Bottom line: heterozygosity has a buffering effect—a variety of proteins allows your body to cope with different environments.

Cool Cats

But does something like a small temperature change really have that much of an effect? For Siamese cats, the answer is yes. With their dark limbs and light-colored bodies, Siamese cats are very distinctive. Their unique coloring is caused by a single temperature sensitive gene. This gene produces a protein that's involved in making a pigment called melanin, the same pigment that makes your skin tan and your hair dark.

Cats with the Siamese color scheme (sometimes called Himalayan when found in other animals) are homozygous for a version of the gene that doesn't work very well at higher temperatures. Since the main part of their bodies is warm, the protein can't perform, and very little pigment is produced. Thus, they have white bodies. The cat's extremities are just a few degrees cooler, but that's enough of a difference for the protein to function properly. Voila! Dark legs, face, ears, and tail.

Owners of outdoor cats are used to dealing with the shenanigans of their feline friends. Gifts of dead birds and lizards are not uncommon. Nor are scratches and missing chunks of fur from fights with other animals. But when your Siamese cat loses fur it may grow back a different color! Missing body fur from feline fisticuffs should grow back white because of the cat's high body temperature. But if the fight happens during the winter, the outdoor temperature can cool the cat's skin enough for the pigment protein to work properly. If so, the fur will grow back dark. I encourage you to get your own Siamese cat and experiment with a razor and an ice-pack. Of course, Siamese cats are so vulnerable to temperature change because they are homozygous for this gene. A heterozygous cat would be buffered against these varying conditions—enzymes that can function at a wider temperature range would ensure consistent pigment production and the cat would be uniformly dark.

Pigment isn't that big of a deal, and the weird coloration of the Siamese cat is actually kind of a cool feature. But this type of temperature sensitivity could just as easily be found in vital proteins involved in muscle growth or oxygen transport. It's one thing to have a white torso; it's another to be unable to get oxygen to your kidneys.

The higher genetic variation provided by heterozygosity can provide a buffering effect against a varying environment. This buffering helps protect an individual from the fickle fluctuations of their surroundings. These fluctuations can be much larger than a temperature change of a few degrees. You can think of the heterosis of sickle cell anemia as an example of heterozygosity buffering against the environment. Here the environmental fluctuation is large: either malaria is present or it isn't. But being heterozygous gives you the ability to cope with either condition. One copy of the sickle cell gene protects you from malaria, and one copy of the normal gene keeps your blood flowing. The heterozygous individual is better prepared than the individual with two sickle cell copies or two normal copies of the gene.

In most cases, being heterozygous for a gene may provide a small advantage or none at all. One gene is just a drop in the bucket— remember you'll be producing forty thousand different types of proteins just today. But what if you were heterozygous for thousands of genes? All those little advantages would start to add up, and you'd end up significantly better at ignoring environmental fluctuations, which means you'd be able to follow your internal blueprint more exactly. And we know how important that is.

HETEROZYGOSITY AND SYMMETRY

Scientists have long predicted that individuals with lower overall heterozygosity will be more affected by their environment and thus less stable during development. Individuals with higher heterozygosity will be better able to follow their internal blueprint which manifests as symmetry (among other ways). With advances in technology, we can now test organisms for different versions of genes and proteins to get an estimate of their overall heterozygosity. This allows us to answer the question: Are more heterozygous individuals more symmetrical?

Yes they are. A relationship between heterozygosity and symmetry has been found in a wide range of animals. Rainbow trout that are more heterozygous have more symmetrical gills, jaws, and fins. Higher heterozygosity was also associated with higher symmetry in the house sparrow, the side-blotched lizard, and in multiple species of bivalves (oysters, clams, etc.).

A study on an unusual fish called the Orange Finned Topminnow (though I prefer its German name, *Orangeflossenkaerpfling*) provides an interesting window into heterozygosity and symmetry. These topminnow bear live young instead of laying eggs, but, even stranger, they can reproduce both sexually and asexually. Young produced sexually are a blend of their parents' genes, but asexual young are clones, exact copies of the mother. By looking at a group of clones, we can see how individuals with the exact same genetic makeup cope in different environments. It's like an identical twin study times one hundred.

The authors of the study examined a group of topminnow clones with very high heterozygosity. They all had the same level since they are all genetically identical. Even though the fish lived in dramatically different environments, they were all highly symmetrical. Their high heterozygosity allowed them to develop stably and consistently despite different external environments. In contrast, the sexually produced fish had varying levels of heterozygosity and varying levels of symmetry. As we have predicted, the more symmetrical fish were the ones that were more heterozygous.

Tamarins are squirrel-sized monkeys that live in the South American rain forest. They are so adorable that you'd think they all have perfectly symmetry, but like all animals, some tamarins are more symmetrical than others. And the tamarins with the least heterozygosity are 200 percent more lopsided than their highly heterozygous peers.

And remember our old friend the cheetah? Last time we talked about how incredibly fast they are. Surprisingly, cheetahs have very low levels of heterozygosity, on average. A study of cheetah symmetry found that they were significantly more asymmetrical than other cat species with higher heterozygosity such as leopards, ocelots, and margays. Imagine how much faster cheetahs could run if higher heterozygosity allowed them to build stronger, quicker, more symmetrical bodies.

Your Body's Most Wanted

We are constantly under attack. Enemies we cannot even see are trying to infiltrate our bodies and consume us from within. Luckily, we have evolved strong defenses against these invaders and can usually fight them off with little more than a scratchy throat and runny nose to show for it. But like any epic battle, it's crucial for your body to know who's on your side. You need to unleash your white blood cells, with their lethal cargo of digestive enzymes, on foreign bacteria, not your own liver by mistake. In order to make sure we're targeting the bad guys, we have a sophisticated system in place to handle self- versus non-self-recognition.

A key component of this system is the major histocompatibility complex (MHC). Also known as the HLA in humans, the MHC is a set of multiple genes that help identify germs so that your immune system can attack them. You can think of the MHC as the set of wanted posters at the post office. Without them the enforcers (police/white blood cells) wouldn't know who to go after.

Because there are multiple genes involved and because there are many potential bad guys to be recognized, the MHC has enormous potential for genetic variation. This makes it a great candidate for looking at heterozygosity. Having high heterozygosity for the genes that comprise the MHC can have two advantages: 1) More wanted posters means a greater ability to identify and stop germs which means better disease resistance; and 2) MHC heterozygosity may be correlated with overall heterozygosity, which would mean better symmetry and health in general.

MHC Heterozygosity and Disease

Next time you're pouring milk onto your Cheerios, take a moment to think about Louis Pasteur. An early pioneer in the field of germ theory, Pasteur helped prove that diseases were caused by tiny organisms called germs. Heating could kill these germs and prevent disease, which is why your milk is pasteurized. He also studied how vaccinations containing weakened germs could provide immunity against diseases. Pasteur developed the first rabies vaccine and in 1885 gave it to a nine-year-old boy that had been attacked by a rabid dog. This was a high risk move for Pasteur. He wasn't a licensed physician and could easily have been arrested for providing treatment, especially since the treatment involved injecting the boy with the very disease they were hoping to prevent! The gamble paid off. The boy didn't get rabies, and Pasteur was hailed as a hero.

It's somewhat fitting then that in 1983, the Institut Pasteur (that Louis founded in 1887) was the first to identify HIV as the virus that causes AIDS. A vaccine still eludes us, but patients infected with HIV do not develop AIDS immediately. Their bodies are able to resist the onset of AIDS for a while. In theory, the stronger the immune system, the longer it can resist the progression of the disease. It turns out that MHC heterozygosity provides a substantial benefit. A study that tracked AIDS onset in patients with different levels of MHC heterozygosity found that being more heterozygous significantly extended survival. Six years after testing positive for HIV, 70 percent of the highly homozygous group had progressed to AIDS. For most heterozygous group it was only 20 percent.

If you live in West Africa, by the time you become an adult you have a 90 percent chance of being infected with hepatitis B. Obviously resistance to hepatitis would be very advantageous in this environment. A study of Gambians infected with hepatitis B found that some people are far more likely to resist the disease. This resistance was linked to MHC heterozygosity.

The increased ability of MHC heterozygous individuals to fight off infectious diseases has also been seen in the laboratory. Heterozygote advantages have been seen in mice infected with fungus; mice simultaneously infected with *Salmonella* and a neurologically damaging virus; and mice simultaneously infected with three different strains of *Salmonella* and one of *Listeria*. They're really not pulling any punches with these mice. Even when simultaneously infected with four different diseases, the mice with high MHC heterozygosity were more likely to clear the infection, more likely to reproduce, and showed more growth. But when the researchers looked at the control group, which hadn't been infected with anything, they found that mice with high MHC heterozygosity showed more growth there too.

This is a very interesting finding. It's not surprising that MHC heterozygosity helped those mice that were subjected to infections, but why did it help noninfected mice? The researchers speculate that MHC heterozygosity could be associated with overall heterozygosity. If so, the mice with high genetic variation in their MHC also have a greater variety of other genes. This means a greater ability to buffer against environmental fluctuations and a more symmetrical, stronger, faster growing body.

Growing Up Heterozygous

This positive relationship between heterozygosity and growth has been observed in many species. The white-tailed deer, a graceful, fast moving creature, is Wisconsin's official State Wildlife Animal. Some may be more graceful and fast moving than others, however. Females with higher levels of overall heterozygosity are bigger and more fertile than their more homozygous girlfriends. Since the mother's weight is correlated with her baby's, heterozygous females can have bigger, healthier offspring. Heterozygous fetuses were also found to grow faster. Finally, heterozygosity was positively related to antler size in males. Antlers are a secondary sexual trait for male deer and are an important part of deer attraction. Female deer may claim that antler size doesn't matter, but studies show that well-endowed males are better at attracting mates.

Human babies also benefit from heterozygosity. Babies in Rome and New Haven were weighed and tested for heterozygosity. The lower weight babies were significantly more homozygous than those born in the normal weight range.

Tiger salamanders grow faster if they have greater genetic diversity, as do various species of marine bivalves. Heterozygous Pacific oysters weigh more. Blue mussels showed the same effect—the less heterozygous mussels grow more slowly and also show less consistency in their growth rate. This makes sense. The more heterozygous mussels produce a variety of proteins and can handle environmental fluctuations. Thus, they can grow at a fast, constant rate. The mussels that don't have the advantage of this type of buffering are at the mercy of their surroundings and show a wide range of growth rates.

This relationship between heterozygosity and growth rate is also well documented in plants. Heterozygous quaking aspen grow faster, and various species of pine show a correlation between heterozygosity and trunk diameter, root length, and cone production.

Wild animals aren't the only ones whose growth is influenced by genetic variation. Domestic sheep that are heterozygous for a *single* gene grow 10 percent faster than sheep with two copies of the same gene. This gene is responsible for an enzyme that clearly plays an important role in growth and development. Pigs showed a similar relationship—more heterozygous animals gained weight more quickly, but, interestingly, they did so while eating less food. Not only were the genetically diverse pigs able to convert food to body mass faster, but also they were able to do it more efficiently. This isn't surprising. Since they are producing a variety of proteins and thus widening the "sweet spot" at which their enzymes function, the pigs are able to operate closer to peak efficiency. With the right genetic makeup, even pigs can be efficient animals.

So can oysters. A laboratory measured oxygen consumption in oysters under normal and stressful conditions. (How do you stress out an oyster? By placing it in a high-temperature, high-salt environment.) Since animals burn oxygen when doing work, oxygen is a good measure of how much energy is being expended. Under the same stressful conditions, homozygous oysters required 2.5 times as much oxygen as heterozygous ones. The implications for athletics are staggering! Humans may not show as much of a range as oysters, but the increased efficiency of heterozygous athletes should confer a huge advantage.

Oxygen consumption is critical in sports. To release energy for movement, we need oxygen. During exercise our muscles get tired because we can't shuttle oxygen to them fast enough. Athletes are desperate to increase their ability to get oxygen to their muscles. Since red blood cells transport oxygen, athletes often try to increase the number of red blood cells in their blood. There are legal ways to do this, like training at high altitude. And illegal ways, like blood doping, which involves injecting additional red blood cells or taking drugs to boost their production.

If an athlete's muscles can work harder with less oxygen, he can do more with the same number of red blood cells. And oxygen consumption is just one example. There are many aspects of endurance that can benefit from increased efficiency. Intense exercise is mostly fueled by breaking down glycogen. We each store about 1,500 to 2,000 calories of glycogen in our muscles and liver. But strenuous exercise can burn through that in a few hours. If your glycogen reserves are depleted, you are very suddenly in pretty bad shape, experiencing weakness, dizziness, even hallucinations. Cyclists refer to this as "bonking"; runners call it "hitting the wall." Heterozygous individuals can burn their glycogen more efficiently and can do more exercise before their reserves are depleted. More work, less bonk.

Of course you can't change your genes. At least not yet. Heterozygous individuals naturally benefit from increased efficiency, but I wonder if future generations of athletes will cheat by somehow artificially mimicking protein diversity.

Knowing that heterozygosity is associated with efficiency, it's no surprise that symmetrical people score so high in athletic ability. Their genetic diversity provides buffering which allows them to construct a nice, symmetrical body. It also allows them to operate that body more efficiently, especially when pushed to the limits of physical endurance.

WHAM, BAM, THANK YOU CLAM

Fast and efficient growth is important for many animals, but for bivalves it's directly linked with mating success. Tall men often do well with the ladies, and the same can be said for clams. But the clam dating scene is pretty boring. The thing about clams is that they're not very good at moving around, so the male and female are rarely very close to each other. Like any long distance relationship, clam love requires serious commitment and investment, not in flowers or plane tickets, but in sperm.

Since a male clam can't seek out females, he does the next best thing—he releases his sperm into the water and hopes that some of them will reach a receptive, waiting female clam. Not exactly the makings of a romantic comedy, but clams are still around so it must be working. The problem with this strategy is that it's literally a shot in the dark. What if your sperm don't reach any females? No baby clams for you. There's only one thing you can do to improve your odds: make more sperm. It's a numbers game and on average, the more sperm our clam bachelor can release, the more offspring he will have. Being a highly heterozygous clam gives you a big advantage. The faster you grow, the bigger you get, the more sperm you can make, the better your chances of fatherhood.

Clams aren't the only ones cashing in their genetic variation at the baby bank. Macaques are Old World Monkeys that, while cute, are not as cuddly as tamarins. Possibly the most famous macaque is Natasha who lives in the Safari Park Zoo in Israel. Natasha came down with a serious stomach flu in 2004 and nearly died. She survived, but after she recovered, she would only walk on her hind legs like a human. (Macaques can walk erect, but usually move around on all fours.) This led to widespread media coverage and a lot of headlines describing her as "the missing link." One commentator even speculated that Natasha was protesting the stalled Middle East peace process.

Personally, I think that Natasha was saying, "Where are all the heterozygous males?" In the wild, Macaque males with higher levels of MHC heterozygosity have significantly greater reproductive success. Heterozygous males sire an average of 2.32 baby macaques versus only 1.52 for their homozygous peers! And like symmetrical men, the heterozygous Macaques start mating earlier. A big part of this advantage may come from being healthier. Part of this study tracked captive macaques. Over the course of the study, several males had to be removed for medical treatment for a severe parasite infestation. All of the males who required treatment were from the highly homozygous group.

SHINY, HAPPY BUTTERFLIES

Happily for butterflies, their dating scene is a lot more exciting than that of clams. Butterflies are extremely mobile, especially during mating season when males travel far and wide looking for interested females. The male butterflies have to fly and court in hot and cold weather, making efficient flying in various temperatures highly desirable. And in fact, male sulfur butterflies that are heterozygous for genes involved in muscle metabolism have the greatest reproductive success.

I suspect that high levels of heterozygosity may help male sulfur butterflies in other ways too. After all, at my high school, the guys who won "Most Athletic" were very popular, but so were the guys who won "Best Dressed." My friend Ashley refers to men that are physically her type as "shiny" because she'll walk into a room and they'll immediately catch her eye. Ashley would make a good sulfur butterfly.

They all look bright yellow to me, but unlike butterflies, I can't see ultraviolet (UV) light. It turns out that some males reflect a lot more UV light than others—they are literally shiny. And just like Ashley, the female butterflies can't resist. One study looked at males that got lucky versus those that didn't. The successful males were twice as shiny. Shiny wings are associated with youth and virility in sulfur butterflies, so it's not surprising that females find them attractive. Which males are the shiniest? We already know that heterozygous butterflies are stronger, more efficient flyers. Less energy wasted and more access to food means more available energy to spend on shiny wings. I'd bet that the males with the greatest genetic variation can grow the brightest wings and maintain their shine the longest.

I don't have to guess when it comes to spotless starlings. These Mediterranean birds are pretty plain looking. Both sexes are black, but the males have some purple plumage and their throat feathers are significantly longer. In fact, one study correctly identified the sex of adult spotless starlings 100 percent of the time using only the length of their throat feathers. These long throat feathers are a secondary sexual trait that males use during the mating season. To woo females, male spotless starlings sing loudly with their throat feathers erect. It's very romantic. A study of spotless starlings in Spain found the most heterozygous males had the longest throat feathers.

A similar study found that European minnows with high levels of heterozygosity had brighter red abdomens. Which, if you're a minnow, is quite sexy.

But What Do the T-shirts Say?

Starlings and minnows are cute, but I'm most interested in studies that apply directly to people. We have all sorts of interesting secondary sexual characteristics that influence our choice of mates. One that often gets overlooked is scent. In the last chapter, I told you that women preferred the smell of T-shirts worn by symmetrical men. Scent can be an important way to get information about a potential mate. People often seem surprised that humans use scent in this way, but they shouldn't be. After all, perfumes have been used in various forms for thousands of years and are currently a multibillion dollar industry.

One needs look no further than perfume ads to see the relationship between scent and sex. Whether it's J. Lo in the shower, Britney Spears in lingerie, or Kate Moss wearing nothing but a string of pearls, perfume ads are all about sex. When Coco Chanel, who created legendary Chanel No. 5, was asked where a woman should use perfume, she responded, "Wherever one wants to be kissed."

But if we use scent to evaluate members of the opposite sex, exactly what are we trying to sniff out? One way in which we may get information about potential mate quality is via the MHC. Remember, the MHC helps us separate self from non-self, so it's reasonable that it can also evaluate non-selves. In fact, instead of covering up our natural scent it's likely that we use perfumes to amplify it. In one study, people with similar MHC types preferred the same perfumes when asked, "Would you like to smell that on yourself?" Their preferences hadn't changed when the experiment was repeated two years later. Perfume ingredients may help us "advertise" our MHC to others. Since our MHCs don't change, our taste in perfume doesn't either. This explains why perfumes can remain popular for long periods of time. Chanel's No. 5 (named after Coco's favorite number) has remained a big seller since its introduction in 1921. Indeed, many of the ingredients in today's fragrances are the same ones used thousands of years ago in China, India, and Egypt.

We are all walking billboards advertising our MHC for potential mates. So who is getting the most business? For the answer to that we must look to the smelly T-shirt experts at the University of New Mexico. They examined this specific question in another intriguing T-shirt sniffing experiment.

The experimental design was similar to the one we discussed in the last chapter. Participants slept in brand-new T-shirts for two consecutive nights and avoided smell contamination by not smoking, not eating garlic, and not using deodorant. The T-shirt wearers were also measured for symmetry, and a blood sample was drawn for MHC genetic analysis. A separate set of opposite sex participants smelled each of the T-shirts and rated them 1–10 for pleasantness, sexiness, and intensity.

As before, the women sniffers rated the T-shirts worn by the symmetrical men as smelling more attractive. But symmetry wasn't the only variable associated with a high smell score. T-shirts worn by men with high levels of MHC heterozygosity were also rated more highly by the women. Bottom line: genetic variation is sexy.

Winning the Heterozygosity Jackpot

All right, what have we learned? It's better to be a butterfly than a clam. Actually it's better to be pretty much anything than a clam. Also, heterozygosity makes you smell good, look sexy, grow fast, resist infections, and do it all more efficiently than your more homozygous buddies. This genetic variation is also the key to that all important trait: symmetry.

We've outlined the fantastic advantages that symmetrical people possess. We now know that this higher symmetry is a direct result of high heterozygosity. A greater variety of genes means a greater ability to compensate for changing environmental conditions and results in a more symmetrical, healthier, and more attractive body.

Perhaps now your focus has shifted from "How do I become symmetrical?" to "How do I become heterozygous?" We know that heterozygosity is the reason some people are more symmetrical, but why would some individuals have more heterozygous genes than others? For the answer, read on.

4

Assembling the Crew

THE BENEFIT OF
INTERRACIAL MARRIAGE

IN THE LAST chapter I was pretty down on clams and their less than exciting sex lives. But if clams are the station wagons of sex, then bedbugs are a Monster Truck Demolition Derby. In general, sex in the insect world can get pretty strange. Male insects can be relentless in their attempts to be the only one to mate with a particular female. Insect penises in some species come equipped with hooks and barbs. The male stays attached to the female for as long as possible to increase the chances that her eggs are fertilized by his sperm. Sometimes for several days. This strategy is known as mate guarding, and it can reach ridiculous extremes. In some species of frogs, the male remains mounted for up to six months!

That's an impressive display of commitment, but some male insects take things even further. They will break off their penis inside the female. Ouch. It's the ultimate sacrifice, but it helps prevent other males from mating with her afterward. Similarly, but less painfully, sometimes male insects will try to seal up the female's vagina with a glue or cement to block future suitors. Not surprisingly, males in these species have evolved shovel-shaped penises to try to dislodge these "copulatory plugs."

But even among other insects, the sex life of a bedbug is pretty dramatic. One entomologist described the love lives of bedbugs as making "Sodom as pure as the Vatican." I outlined how male insects have turned the vagina into a veritable battlefield. Possibly as a result of these tactics, bedbugs have evolved to bypass the vagina altogether. Bedbugs mate by a process called "traumatic copulation"—the male stabs the female in the abdomen with his penis and ejaculates sperm directly into her body cavity. Hey, I warned you it was strange.

The sperm swim around inside the female until they find a special storage gland and settle in. The female will keep the sperm there until she is ready to fertilize her eggs. Female bedbugs have even evolved a special layer of tissue called the Organ of Berlese that helps them survive being stabbed in the abdomen. But wait, it gets stranger.

In a surprising turn, scientists have observed male bedbugs stabbing other males! It's comforting that females aren't the only ones getting punctured, but why? At first it was assumed to be an accident—stab-happy males or a case of mistaken identity, but it turns out to be a deliberate strategy. Just as with females, after copulation the stabber's sperm swim around and find the reproductive tract. In another male, they set up camp in the vas deferens (sperm duct) and wait for him to inseminate a female. When he does, they hitch a ride. It's not bad enough that this poor guy got stabbed; now he could be fertilizing a female's eggs with another guy's sperm. His reproductive system has been hijacked!

I warned you it was bizarre, but the antics of bedbugs make perfect evolutionary sense. Bedbugs aren't cruel or perverted; they're just trying to maximize their reproductive success. Piggybacking your sperm inside another male is a good strategy because it will increase your total number of offspring. Therefore, males that employ this strategy have more offspring, and the trait is passed down. Males that didn't do it were at a reproductive disadvantage and died out. It would appear that, in this case, nice bedbugs finish last. I find it fascinating that the same gradual, yet powerful forces of natural selection that led to the extraordinary speed and grace of the cheetah, have also given us homosexual bedbug rape.

WHY DO WE HAVE SEX?

Bedbugs make the human dating scene look absolutely tame. Not that I'm complaining—I'm quite happy that getting lucky does not entail anyone getting stabbed in the gut. Plus, we are not completely humdrum in our amorous pursuits. I'm certain that you personally have witnessed or been involved in some crazy misadventures undertaken in the name of love or sex. So why do we do it? What makes sex so desirable that bedbugs are willing to stab and be stabbed to get it? Why did my friend Alan fly a girl he had never met to Las Vegas because she sounded cute over the phone? Why is sex so compelling?

This leads to an even more fundamental question. It's one thing to ask why we are culturally or individually obsessed with sex, but why do we have sex at all?

I've given talks at universities in different parts of the United States and I always pose this question. It's a little dangerous to ask college students about sex. There's always a stunned silence followed by some tittering in the back. Then some glances are exchanged that seem to say, "Why do we have sex? Is he kidding?" I've gotten a few humorous responses, but eventually someone will step up and sheepishly deliver what they assume is the biologically correct answer I'm looking for: "To reproduce." And each time they're so surprised when I tell them they're wrong.

Don't misunderstand me, I'm not trying to rewrite the birds and the bees here—sex *is* how we reproduce, but that's not the whole story. Sex is only one way that reproduction can happen, and it has some disadvantages compared with other methods. The first life forms that evolved on our planet reproduced asexually, and many still do. Most bacteria do not reproduce sexually, when they are ready to have a family they just divide—producing two identical daughter cells. Asexual reproduction isn't that exciting, but it does have two significant advantages over sexual reproduction:

1. It's convenient. No partner necessary means you can do it anytime, anywhere. It's not necessary to hope there will be a willing and able mate at the right place and time. No need to buy anyone a plane ticket. No need to expose yourself to predators or sexually transmitted diseases.
2. You get to pass along all of your genes. The goal of reproduction is to get your genes into the next generation. With sexual reproduction you have to share the wealth—only half of your genes make it into your offspring. Asexual reproduction, on the other hand, results in identical clones with all of your genes.

But despite these two points in favor of asexual reproduction, sex is extremely common. Why? Because it has one huge advantage. Sex doesn't just give us reproduction; it gives us reproduction with variation. And that variation is critical to succeed in a changing world.

I'm Thinking about Starting a Family

What if we didn't reproduce sexually? Cloning technology is getting more advanced everyday. Let's say I decide to take advantage of these new opportunities and reproduce asexually. Life would be simpler. I would have no need for a mate; I could do everything myself. When I felt like I was ready to start a family, I could just create some clones of myself. A couple of adorable mini-Alons, identical to me in every respect. And when my Alon-babies grow up, they could have some Alons of their own. So far this plan is sounding pretty good. In a few generations there will be hundreds of Alons, each one carrying all of my genes. I'm swelling with pride just thinking about it. But there's one problem: the weather.

You see, I hate the heat. I just can't handle hot weather, and I fare far better when it's cold. Since my children and grandchildren would be identical genetic copies of me, they would all struggle with heat, too. It's bearable now, but if the global warming indicators are right, it will soon become far more dire. That's bad news for me and all of my Alon-descendants. If it gets hot enough, it could even prove fatal. And if that's the case, it will be fatal for every single one of us. For me, my children, and all of my progeny—because we're all exactly the same. My entire line would be wiped out in one fell swoop.

The tragedy of the Alon Clan highlights the problem with asexual reproduction. Exact copies are vulnerable to the exact same things. They may be adapted for the current environment, but a single change can destroy an entire family. Putting all of your genetic eggs in one basket (or a bunch of identical baskets) is very risky.

But let's say I decide to do things the old-fashioned way. I find a wonderful wife, move to the suburbs, and we have kids together. Each of our kids has half of her genes and half of mine. I don't get to pass along as many genes, so my children are similar to me, but not identical. Each is a unique combination of our genes. They're different from their parents and from each other.

Not quite as impressive as the army of Alons I was imagining earlier, but this variation is important. Especially in a few generations when global warming results in Sahara-like conditions. Some of my descendants may have inherited my heat-intolerance and may perish, but some will be able to tolerate the heat, and so my family and my genes will live on. Sexual reproduction provides variation, which allows us to cope with a changing environment.

And *that* is why we have sex.

YES, WE HAVE NO BANANAS

Maybe the troop of identical Alon-clones sounds a little farfetched, but you don't have to look too hard to find a real world example of this. Actually, you only have to look in your lunchbox. If you're like most Americans, there's a banana in there. The average American eats over 26 pounds of bananas per year. By far the most of any fresh fruit (apples are a distant second at around 16 pounds per year). And why not? Bananas are good. Chiquita claims they are "quite possibly the world's perfect food." That's a little self-congratulatory, but the truth is that bananas are tasty, good for you, and come in their own convenient packaging. They're also clones.

I don't mean for that to sound ominous; it's just a fact. After all, when was the last time you bit into a banana seed? New banana plants aren't grown from seeds; they're grown from shoots cut from another plant. Pretty much all of the bananas that we eat in the United States, Canada, and Europe are the same variety—Cavendish. All Cavendish plants are grown from cuttings from other Cavendish plants. Going all the way back to one plant discovered in Southeast Asia one-hundred years ago. So every Cavendish banana contains exactly the same DNA; they're genetically identical clones.

This uniformity makes for very consistent fruit, but it also makes the banana crop very vulnerable. Just like the Alon family. In this case, the problem isn't heat, it's fungus. We're eating millions of Cavendish bananas every year, but fifty years ago we weren't eating any Cavendish. Fifty years ago there was a different banana in town, the Gros Michel or "Big Mike." If you ask older folks, they may remember the "Big Mike." It was a little larger than the Cavendish and supposedly tasted better.

So what happened? The worldwide Gros Michel crop was devastated by Panama fungus. At first, banana growers tried to outrun the disease by moving their crops to new fields, but eventually they had to give up. The lack of variation in the "Big Mike" meant that if the fungus could destroy one plant it could destroy them all, and that's what happened. In the 1960s, the major banana companies abandoned the Gros Michel and started growing Cavendish instead, because the Cavendish is naturally resistant to the Panama fungus.

Or at least it used to be. In 1992, a new strain of Panama fungus was discovered that attacks Cavendish plants. Since the banana companies didn't learn from the last banana catastrophe, it looks like they may experience another one. Once again the crop is genetically uniform and thus extremely vulnerable. The Cavendish may soon go the way of the "Big Mike." Fruit researchers are already searching for a replacement banana. I suggest you enjoy a banana split while you can; the flavor may soon be very different. By putting all their eggs in one genetic basket, the banana companies have doomed themselves to a cycle of major disasters. Fruits that reproduce sexually are far less vulnerable. The built-in variation of a sexually reproducing species protects it from this type of environmental calamity.

THE ANSWER TO EVERYTHING THUS FAR

If you've been paying attention, this may sound very familiar. "Variation allows us to cope with a changing environment." That's exactly what I said was the great benefit of heterozygosity. In that case I was talking about variation on the genetic level (within an individual); here I'm talking about variation on the level of the population. But the two concepts are similar, and in both cases the source of the variation is the same: sex.

In the last chapter, I went on and on about the benefits of heterozygosity without telling you where it comes from. Well now I am: it comes from sex. When you and your partner have a baby, you each contribute one copy of each gene. If the two copies are different, then the child is heterozygous for that gene. By mixing our genes up sexually, we create genetic variation in our offspring.

Of course in talking about the benefits of heterozygosity and symmetry I made it quite clear that some lucky people have higher levels than others. We're all products of sex, so why are some individuals more heterozygous and thus more symmetrical, attractive, and healthy?

Remember, your parents can give you two different versions of a gene, increasing your overall heterozygosity, or they can both give you the same version of the gene. The way that your parents' genes combine is what determines your genetic makeup, including your level of heterozygosity. The more similar your parents are to each other, the more genes they will have of the same versions. That means they will pass those same versions on to you, and you will be pretty homozygous.

This is the problem with incest. Two incestuous parents are genetically very similar because they are closely related. They are passing a huge number of the same genes onto their kids which means very high levels of homozygosity. This results in a higher chance of recessive genetic disorders (inheriting two defective copies of a gene) and a generally reduced ability to cope with environmental changes (because of low genetic variation).

Incest is bad because the parents are too similar genetically. Parents that differ in their DNA will produce healthier offspring. Mating with nonrelatives is better because there is less genetic similarity. Even better is having kids with someone from another race. The genetic differences here are even greater, which means even higher levels of heterozygosity for the children.

It's taken us a few chapters to get here, but this is the core message of the book: Interracial people are a mix of their parents' very different gene pools. The greater genetic variation that they inherit allows them to follow their DNA blueprint more closely and build stronger, healthier, better-looking, more symmetrical bodies.

BUT I HEARD THAT RACE DOESN'T EXIST

When I give talks on the topic of interracial advantages, my favorite part is the question-and-answer session at the end. It's fun to interact with the audience, and I hear a lot of different questions. But without fail there are two questions that I always get. We'll talk about the second question a little later, but the first one is always something like, "Isn't race just a social construct?" Or, "Isn't it true that there's no such thing as race?" Or, "Isn't there more variation within a race than between races?"

All these questions are asking the same thing: Are the different races really different on a genetic level? I've argued that two parents from the same race will be more similar, and their children will have less genetic diversity as a result. Is that true? Yes, and to see why let's take a look at my family tree.

I have 2 wonderful parents. They each in turn had 2 parents, giving me 4 grandparents. I didn't know them, but each of my grandparents also had 2 parents for a total of 8 great-grandparents. One more step back gives me 16 great-great-grandparents. That's nice, but what's the big deal? Doesn't every family tree look like that? Actually, no.

My friend Tannaz has a slightly different family tree. She has 2 parents and 4 grandparents, but 2 of her grandparents were cousins. Since they were cousins, their parents were siblings. Since their parents were siblings, they shared a set of grandparents. That sentence may be a little confusing, but if you think about one of your cousins, you'll see that between you there are only 3 sets of grandparents because you share one set. Between you and a stranger, there are 4 sets of grandparents. Bottom line: Tannaz only has 14 great-great-grandparents versus my 16.

In western society, first-cousin marriage is generally frowned upon. According to cousincouples.com, it's illegal in twenty-four states. The truth is that first-cousin marriage isn't nearly as bad as close incest because you and your cousin don't share as many genes. Still, I wouldn't recommend it. Advocates of first-cousin marriage are quick to point out that Charles Darwin, the father of evolution, married his first cousin. They don't mention that he blamed that marriage for many of the health problems his children suffered.

It's easy to feel superior to Tannaz and her shrunken family tree, but the truth is that we all have this kind of familial overlap. I don't know of any in the last few generations of my family, but a simple thought experiment shows that there has to be some: Without the overlap caused by incest, the number of my ancestors doubles every generation as we go back in time: 4 grandparents, 8 great-grandparents, and so on. But what if we go back 30 generations? With no overlap I would have to have over 1 billion ancestors. Thirty generations sounds like a lot, but if we assume an average generation time of 25 years, it's only 750 years ago. The problem is that 750 years ago there weren't even 1 billion people on the planet, much less 1 billion of my direct ancestors!

Obviously there has been some overlap in my family. This can happen knowingly, as with Tannaz's grandparents, or unknowingly—would you recognize your seventh cousin? This type of inbreeding reduces the amount of genetic variation, but can be unavoidable, especially in a small isolated group.

WE MARRY OUR OWN

Often groups are forced to inbreed because they are geographically isolated. If no one else is around, you make do with the choices you have. This is the case with Iceland. This arctic island nation is extremely remote (Reykjavik is the northernmost national capital in the world) and inhospitable (Iceland has more land covered by glaciers than all of continental Europe). Not surprisingly, Iceland hasn't drawn many immigrants historically. Therefore, the almost three-hundred thousand Icelanders who currently reside there are, by and large, the descendants of a small group of Viking settlers who arrived during the late ninth and tenth centuries. Geneticists believe that all Icelanders are related to each other to some degree.

Geography isn't the only barrier to intermixing. The Old Testament has a clear policy concerning Jews marrying non-Jews. Deuteronomy 7:3–4 states, "Neither shalt thou make marriages with them; thy daughter thou shalt not give unto his son, nor his daughter shalt thou take unto thy son. For they will turn away thy son from following me, that they may serve other gods: so will the *anger of the LORD be kindled against you, and destroy thee suddenly.*" I added the emphasis, but it hardly needs it. Being threatened with destruction by God's wrath is pretty motivating.

So it's no surprise that historically most Jews have married other Jews. But how much mixing has there been? With modern technology we can compare genes with traditions and see how they match up. One important Jewish tradition is that of the kohanim or priests. In Genesis, God commands that Aaron (Moses' older brother) and all of his sons be the high priests of the people. To this day, all of his male heirs assume this role and often have last names like Cohen or Kahn. Since it is a mark of honor, fathers are proud to pass it along to their sons.

Something else that fathers faithfully pass on to their sons is their Y chromosome. Every male receives an X chromosome from his mother and a Y from his father. He will pass that same Y chromosome along to his sons, if he has any. So, both the kohan title and the Y chromosome should represent an unbroken paternal line. If the traditions have been kept, this line will lead directly back to Aaron. Thus, every kohan today should carry a copy of Aaron's Y chromosome. It's possible that some mutations may have occurred in the intervening centuries, but the Y chromosomes should still be similar enough to identify.

Professor Karl Skorecki and collaborators from Haifa Technion went looking for Aaron's Y chromosome. They collected DNA samples from two hundred male Jews from Israel, the United Kingdom, and North America. When they analyzed the samples, they found that non-kohan Jews had a variety of Y chromosomes, a similar variety to other groups from the Middle East. But Jews who identified themselves as kohanim often had a particular set of genetic markers that indicated a shared Y chromosome. Fifty percent of the kohanim carried these markers, which are now called the Cohen Modal Haplotype. Fifty percent is not a perfect track record, but it is impressive that a large number of kohanim, who over hundreds of generations have spread all over the world, really do come from a single male ancestor.

What about the other 50 percent? It's easy to blame intermarriage and conversion to Judaism, but converts are not allowed to become kohanim. It's possible that some men claimed kohan status (either accidentally or intentionally) when it didn't actually belong to them. Another factor (that any geneticist learns early on) is that we don't live in a perfectly monogamous world and sometimes a person's father is not who he thinks.

As I mentioned, non-kohan Jewish men don't seem to have any special Y chromosome markers that are unique to them. This is probably because of mixing with other groups and the occasional religious conversion. This type of mixing also explains why European, Middle Eastern, and African Jews have such different appearances. But it would appear that for the majority of their history the Jews followed God's prohibition. Other Y chromosome studies have shown that in each generation, less than 1 percent of Jewish women had children with non-Jewish men.

By staying within the group, genetic diversity is reduced, and an ethnic genetic identity emerges. Jews have a different genetic makeup than other groups. The most obvious differences are the genes that cause disease. My father found out the hard way that Gaucher disease is far more common among Jews of European descent than other groups. Most likely at some point in history one Jew (let's call him Manny) experienced a mutation in his glucocerebrosidase gene. This was a significant event—one that would sow the seeds for my father's disease hundreds of years later. An event that ultimately affected the lives of thousands of Jews for many generations to come. But Manny didn't even notice.

Manny still had one good copy, so he was healthy. Like me, he was just a carrier. So Manny lived his life and had some kids. Some of his kids inherited the normal version and some inherited the defective version. But still, no one actually had the disease. In the small, isolated communities where Jews traditionally lived, it would be easy for the gene spread over time. Eventually two carriers of the gene would marry and have kids. Some of those kids would be the first victims of Gaucher disease.

The two parents had both inherited the defective gene from Manny, who had the original mutation. They were both related to him and thus to each other. They may have been distant relatives, but they both had the defective gene that must have come from Manny. (Two independent mutations is far less likely.)

This is why I don't feel superior to Tannaz. My grandparents both passed along the defective Gaucher gene to my father. Both of those genes ultimately came from Manny. So my grandparents were related to each other. They weren't first cousins, but those two genes can only come together through overlap. And that overlap is far more likely with two people of the same ethnic group.

MIXING IT UP

I'm not saying that throughout history people never married outside of their race, just that it was less common than marrying within their race. Depending on geographic and cultural boundaries, different groups experienced different levels of mixing. The Middle East has experienced multiple waves of migration. More and more evidence points to modern humans originating in Africa. Their route out of Africa and into the rest of the world passed through the Middle East. Later Middle Eastern residents continued to migrate into Europe and Asia. These populations became ethnically distinct over time, and some returned to the Middle East, leading to more mixing. Its status as the crossroads of the ancient world has resulted in today's Middle East—one of the most genetically diverse areas of our planet.

Sadly, not all mixing has been voluntary. Slavery is a dark stain on the history of the United States and all nations that participated in it. It's estimated that over approximately four-hundred years, the Atlantic slave trade captured and transported 12 million slaves to the New World, primarily from West Africa. Those that survived the dreaded Middle Passage across the Atlantic could look forward to a life of servitude and often harsh treatment. Today's African Americans are their descendants. But how are African Americans different from Africans?

One significant difference is found in their genes. Over many generations, the African American population has absorbed some European genes. Surely some of this happened through mixed marriages, but the ugly truth is that many slave owners raped their slaves. As I mentioned in Chapter One, various "one drop" rules ensured that the children of these forced unions would be classified as black, despite being half white. The end result is that African Americans possess a significant minority of European genes. The majority of their DNA can be traced to their African roots, but studies have found 6.8 to 22.5 percent of African American genes to be of European origin. This is why African Americans often have lighter skin than modern-day West Africans.

You're What?

Sometimes evidence of mixing can be found in the most unlikely places. This is the case with the Lemba. The Lemba are a tribe located in South Africa and Zimbabwe. Like other groups in southern Africa, the Lemba look African and speak Bantu languages. However, even though they look like their neighbors and speak like their neighbors, the Lemba claim to be different. Specifically, they claim to be Jewish.

As you might expect, this claim has been met with a lot of skepticism. But the Lemba do have some evidence on their side. They have Semitic-sounding surnames and legends detailing their origin in a place called Sena, which may be located in modern Yemen. Lemba tradition forbids the consumption of pork and outlines specific rules for meat preparation similar to the Jewish kosher laws. They also practice ritual circumcision and must marry within their group.

Could they possibly be following the same prohibition against intermarriage written in Deuteronomy?

Recently geneticists tried to solve the riddle of the origin of the Lemba. With permission from Lemba elders, they took DNA samples from males in the various Lemba clans. The Lemba were eager to participate with the hope that it would validate their claims to Jewish heritage. And their hopes were realized.

Analysis of the men's Y chromosomes found two-thirds were of Middle Eastern origin. And just like other Jewish populations, some of the men carried the Cohen Modal Haplotype. Among the Buba, the oldest and most respected of the Lemba clans, over 50 percent of the men had the Cohen Modal Haplotype. The Buba may have achieved their premier status because their patriarchs were members of the Jewish priesthood. This is clear evidence that the Lemba are descended from Jews, at least on their paternal side.

Just as the Y chromosome is inherited paternally, cells' mitochondria are passed on only through the mother. Analysis of Lemba mitochondrial DNA indicates its origin to be African. The Lemba forefathers may have been Jews, but most likely they were marrying native Africans.

However, while mixing did occur, I maintain that in the ancient world, you were far more likely to have children with someone like you than someone different from you. This is what led to the development of ethnic groups in the first place. Groups that are still identifiable today. Even groups like the Lemba, which originated via mixing, have remained relatively isolated since their inception. This type of isolation leads to the overlap I talked about earlier. And that overlap leads to a loss of genetic diversity.

An extreme case of this is the Samaritans. Most people have heard the New Testament fable of the Good Samaritan. In this story, a Jewish traveler is attacked, robbed, and left for dead. Later a priest travels down the same road, but offers no assistance. Another traveler comes along, but also ignores the wounded man. Finally, a Samaritan recognizes the traveler's serious condition and offers both medical and financial help. Usually the moral conclusion is to help your fellow man.

Most people are unaware of the political context of this story. Specifically that Jews and Samaritans hated each others' guts. In this story, Jesus wasn't preaching about simple generosity, but about ancient race relations! Samaritanism is similar to Judaism in many ways, but both reject the other's interpretation. So what do you do when you are surrounded by people who hate you so much that it's a big deal when Jesus tells a story showing you in a good light? You keep to yourself.

Today, the Samaritans are a very small and very isolated community in Israel and the Palestinian territories. Only about 650 members remain, all descended from four families. As a result, the Samaritans are one of the most inbred groups in the world. More than 80 percent of Samaritan marriages are between either first or second cousins. The situation has become so dire that all Samaritan marriages must be first approved by a geneticist at Israel's Tel HaShomer Hospital. In an attempt to remedy the situation, Samaritans have recently decided to allow non-Samaritan women into the community. However, because Samaritans have strict rules concerning women (e.g., women must be isolated during menstruation), female converts are few.

This is not just an issue in small communities. In many parts of the world, cousin marriage is encouraged. In parts of Arabia and Africa, cousin marriages are often seen as the ideal arrangement. Some West Africans have a saying, "Cousins are made for cousins." One survey in Bagdad in 1986 found that 46 percent of couples were first or second cousins. In the United States, the number is more like .2 percent. This means Iraqis are about 230 times more likely to marry their cousins than Americans.

Both physical and cultural constraints can prevent mixing and lessen genetic diversity over time. Interracial marriage is a way to restore this lost diversity. When two parents of different ethnic backgrounds and genetic stock come together, their children are a highly diverse mix of their DNA. Thus, these children have the high heterozygosity that we've established is one of the main keys to health, beauty, and vigor. But why am I the only one saying this? Why is the conventional wisdom that races are all the same?

THE EMPEROR HAS NO RACE

I think the main reason is that science and race have a very ugly history. Historically, race science was often involved in proving the superiority of one race (the one doing the science, of course) over another. In this way, Europeans "proved" themselves to be superior in intelligence, strength, even morality. They then used this superiority as justification for things like imperialism, slavery, social Darwinism, and eugenics. The problem is that the study of human diversity became inevitably linked to these racist practices. They are completely separate, but I think the fear is that one cannot explore the differences among people without determining that one type of person is better than another. This leads to a dilemma: If we acknowledge that ethnic groups are truly different, then we open up a Pandora's Box of racial ranking. But if we deny that there are real, genetic differences between groups, then we don't raise any uncomfortable questions. It's much easier and safer to deny that race exists.

Never mind that different races are known to suffer from different genetic disorders; or that members of different races can respond very differently to the same medication; or that the FDA recently approved BiDil, a heart medication specifically for African Americans; or that 95 percent of Asian Americans are lactose-intolerant compared to about 10 percent of European Americans; or any other evidence that ethnic groups have genetic differences. Life is simpler if we pretend that we're all exactly the same.

Of course the majority of scientists are honest and reasonably objective. They don't lie or falsify data to perpetuate the illusion of no race. But I think the fear of racism has created a strange kind of doublethink in the scientific community. You can study race; you can examine the differences between groups; you can publish your results; you can explore racial differences openly and objectively, as long as in your concluding paragraph you say that race doesn't exist. I'm completely serious.

I saw a great example of this in a recent issue of *Scientific American*. I was excited when I pulled the magazine out of my mailbox because the cover story was right up my alley. DOES RACE EXIST? was emblazoned across the front. This is an area I'm obviously interested in, so I quickly flipped through to the story.

It was a smart and well-written article by two experts in the field. One of the main sections of the article dealt with trying to identify people's race using information from their DNA. As DNA is passed along through many generations, it changes because of mutations. As I have argued, since most groups have been relatively isolated since they separated from each other, people from the same group should share many of these mutations. Certain mutations or sets of mutations can be connected to specific groups.

This is exactly what scientists have tested. One of the authors of the article, Michael J. Bamshad of the University of Utah School of Medicine, looked at a specific kind of mutation called an *Alu*. These are short pieces of DNA that occasionally copy themselves to a new location in one's DNA. They generally don't affect genes so they don't have a positive or negative effect. Since they're neutral, they just hang around and get passed along to one's children. Thus if two people share an *Alu* in the same place, they must have received it from a common ancestor. Bamshad and his colleagues were able to assign people to one of four groups: Europeans, East Asians, and two different groups of Africans, based only on their *Alu* sequences. By examining sixty different *Alus*, they were able to sort the people into the appropriate category 90 percent of the time. When they increased the number to one-hundred *Alus*, they approached 100 percent accuracy.

Another study recounted in the article looked at a different type of DNA sequence, and showed similar results in sorting people into five general geographic origins: sub-Saharan Africa, Europe and West Asia, East Asia, Oceana (New Guinea and Melanesia), and the Americas. This study also attempted to sort people into finer-grained subgroups with a high success rate. I was intrigued by these findings and also impressed with *Scientific American*. Here a popular magazine was taking a stand—yes, ethnic groups are different, and we can see these differences on a purely genetic level.

A company named DNAPrint Genomics offers a similar service to law enforcement agencies. By analyzing DNA left at the scene of a crime for seventy-three genetic markers, they can accurately determine a person's race. They can also estimate the percentages of multiple backgrounds in a suspect of mixed race.

Later I decided to see what else this issue of *Scientific American* had to offer. As I was scanning the table of contents, I noticed the entry for the cover story (hard to miss, it was highlighted in red): "Does Race Exist?" and then the first line of the article summary, "From a purely genetic standpoint, no." Huh? According to the article, we can determine an individual's ethnic background with an extremely high rate of accuracy using only genetic information, yet somehow race doesn't exist "from a purely genetic standpoint." It's hard to see the logic here, but this way Pandora's Box remains reassuringly shut.

Perhaps the best example of this attitude is a massive tome called *The History and Geography of Human Genes*. L. Luca Cavalli-Sforza is professor of genetics at Stanford University and a towering figure in the field of human population genetics. This book, which he wrote with two coauthors, is the bible of human diversity; except I think it may be longer than the actual Bible. The original version weighed in at over one-thousand pages and 7.5 pounds. It costs $250, but for your money you get a huge amount of genetic data (over 110 genes mapped in almost two-thousand human populations) which the authors use to reconstruct human migrations and create a racial "family tree" depicting how various ethnic groups are related to each other.

It's a dense, but fascinating book. But to see how it fits into the doublethink of race science politics, you don't even need to open it. On the back cover, the book is lauded by various publications. *Time* magazine writes that it proves "racial differences are only skin deep." This sentiment is a common one. Jared Diamond, a famous author in his own right, says that Cavalli-Sforza's research is "demolishing scientists' attempts to classify human populations into races in the same way that they classify birds and other species into races." The *New York Times* refers to a more recent Cavalli-Sforza book as, "dismantling the idea of race."

To see how false these assertions are, one needs only to flip over *The History and Geography of Human Genes* and look at the front cover. All the research in this weighty volume is distilled into one image—a map depicting the human genetic geography of the world. And this map is on the front cover of the book. But the map doesn't "dismantle" any ideas about anything. Sub-Saharan Africa is in one color (yellow), Europe another (green), and Asia yet another (blue). The Americas are purple and Australia is a deep red. I think that Steve Sailer, founder of the Human Biodiversity Institute, put it best when he wrote, "Basically, all his number-crunching has produced a map that looks about like what you'd get if you gave Strom Thurmond a paper napkin and a box of crayons and had him draw a racial map of the world."

Sailer's colorful assessment is dead-on. The only thing surprising about Cavalli-Sforza's map is how much it disputes what everyone else says about his research. The map clearly says that human differences are real and that they follow the general pattern that any of us would predict.

OPENING PANDORA'S BOX

There, I've said it: race is a real concept. I've opened the Pandora's Box of scary questions. But it turns out that they're not that scary after all. Here are the most obvious ones.

Are Races Completely Distinct?

No. I've already given some examples of historical mixing. Today's world is a much smaller place, making mixing that much easier. Races are just extended families. Sometimes people marry into other families. Plus we all came from the same origin about one hundred thousand years ago. I've emphasized our differences in this chapter, but we're still all humans, descended from the same African ancestors. Along the way we've become separated into vague extended families, but we're still more similar than different. Of course, those differences are still significant and important.

How many races are there?

This is related to the last question. The truth is that I have no idea. Those who classify ethnic groups are usually lumpers, who put everyone into a small number of big groups (the four or five general races), or splitters, who divide people into a higher number of smaller groups (Cavalli-Sforza analyzes forty-two different groups). Personally, I think of race as more of a continuum. I believe that we have differences, but because of our common origins and occasional intermarriage, they are not easily categorized. Still, we know that some people are more similar to us and some more different, regardless of labels. If you and your spouse both descended from ancestors that lived in the same small Italian village, you are probably genetically similar. Nothing wrong with that, but your kids' level of heterozygosity will reflect that. If your family comes from that Italian village, but your spouse's roots are in Poland, I'd expect higher heterozygosity in your children, even though most people would assign you both the same label, be it Caucasian or European or white. If you mix your Italian DNA with someone of Mongolian extraction, even higher heterozygosity would result.

Is One Race Superior?

No. The idea that one race was chosen by some higher power to dominate the others is ridiculous. Races can be different without falling into some kind of hierarchy. Granny Smith apples are different from Red Delicious, but both are tasty and any attempt to rank them is purely arbitrary.

Are Certain Races Better at Certain Things?

I know I said that Pandora's Box wasn't that scary, but even I'm afraid of this question. I really don't know. This question is dangerous because if the answer is yes, it's very easy to fall into the trap of discrimination. Even unintentionally. There is so much diversity within ethnic groups that it's quite possible the answer is no. If people within one race go from one end of the spectrum to the other, it's hard to imagine one group having a totally different set of skills than another. Still, many group differences have been documented. Sometimes they can be written off as cultural differences, but it's hard to look at something like the way East Africans overwhelmingly dominate distance running, and remain skeptical. I don't know this answer, but I also don't want to dwell on it. I think the most interesting thing about our differences is how powerful they can be when brought together. I find that this question is almost always divisive; the goal of this book is to encourage unity.

THE OTHER QUESTION

Earlier I mentioned that without fail I am asked two questions every time I speak on interracial advantages. The first was, "Is race a real concept?" Once I've convinced the audience that race is real and the benefits of mixing are significant, someone will ask, "So which is the best combination of races?"

Of course every individual is unique, and the process of human development is incredibly complex. We're talking about the complicated interactions of millions of genetic and environmental variables. There is no way to predict how any given scenario will play out. And there are no guarantees. My friend Mike is biracial, yet he suffers from a genetic disorder. And certainly there are many non-biracial people who are beautiful and healthy.

The previous paragraph is 100 percent true. It's also 100 percent unsatisfying. So with all of those caveats in place, I'll try to give a more concrete answer. Heterozygosity is strongly associated with attractiveness and vigor. I have said that the higher one's heterozygosity is, the better. Which individuals will have the highest levels of heterozygosity? The ones whose parents are the most different. So really this question is asking, "Which races are the most different?"

Modern humans originated in sub-Saharan Africa. Some of these humans left Africa and migrated throughout the world. Thus, all non-Africans are descended from a subgroup of the original African population. Therefore, Africans are genetically older and more diverse than other modern-day groups. According to Cavalli-Sforza, "The most important difference in the human gene pool is clearly that between Africans and non-Africans." So, if we're talking generally, I would predict that the combination that would yield the highest heterozygosity would be that of an African and a non-African.

But we can get even more specific. In *The History and Geography of Human Genes*, Cavalli-Sforza and his coauthors analyze forty-two different ethnic groups to try and understand how they are related to each other. To do this they calculate the genetic distance between each group and every other group. The more similar the DNA of two groups is, the smaller their genetic distance. Therefore, the pair with the highest genetic distance is the most removed historically and the most different genetically. The children of this pair would have the highest level of heterozygosity on average.

According to Cavalli-Sforza's research, that highly mismatched pair is Bantu (of Southern Africa) and Eskimo (of Alaska and the far north of North America). These two groups live literally on opposite sides of the world, in two different hemispheres. This geographic divide is mirrored in their DNA. There was a time when a Bantu-Eskimo marriage was impossible; each group didn't even know of the other's existence. Today the world is a much smaller, more integrated place. I don't know of any Bantu-Eskimo love connections, but I'm sure there are some out there.

Remember though, there are no guarantees. I'm certainly not

advocating some kind of race of Bantu-Eskimo supermen. I do think that Bantu-Eskimo children would have very healthy levels of genetic diversity. That's a powerful factor to have on one's side, but it's still just one factor.

So far we have laid out all of the pieces:

☞ Symmetry is associated with increased vigor and attractiveness.
☛ High heterozygosity leads to more symmetry.
☞ Interracial mating produces children with high heterozygosity.

In the next chapter, we put this theory to the test.

5

Inspecting the Towers

PUTTING THE THEORY TO THE TEST

MUTINY IS SERIOUS business. On April 28, 1789, the HMS *Bounty* was sailing home with a cargo of over one thousand Tahitian breadfruit plants. They were to be transplanted to the West Indies to provide cheap food for slaves. The plants would never arrive, however, because on that morning, First Lieutenant Fletcher Christian led a mutiny that successfully took control of the ship. Captain William Bligh and eighteen men loyal to him were set adrift in a twenty-three-foot launch.

The story of the *Bounty* has been told many times in many different forms. There are hundreds of books on the mutiny, including one written by William Bligh himself. At least four separate film versions have been made throughout the years, each featuring one of the biggest stars of his generation: Errol Flynn, Clark Gable, Marlon Brando, and Anthony Hopkins. Many stories have focused on Captain Bligh's leadership in order to determine if the mutiny was justified. Others focused on Bligh's odyssey home, a trip of over 3,600 miles in what was essentially a rowboat; remarkably he and all but one of his men survived.

There is a rather unsatisfying conclusion to the story. Bligh survived and led a second voyage to Tahiti. This expedition was incident free and he successfully transported over two thousand breadfruit plants to the West Indies. However, all of Bligh's hard work was for naught—the slaves there refused to eat the breadfruit.

But what of the mutineers? After a stop on Tahiti, Christian, the lead mutineer, took eight of his men and eleven Tahitian women and started a new colony on the remote island of Pitcairn. This was an excellent choice of hideout because Pitcairn was in the wrong place on the Royal Navy's maps. It would not be rediscovered until 1808. Basically, Christian and his men lived in a remote, tropical paradise with a bunch of young Tahitian women, and no one could find them. Not a bad ending.

As I discussed in Chapter One, it was commonly accepted at that time that mixed marriages were unnatural, and the offspring would be physically and intellectually inferior. The inhabitants of Pitcairn found quite the opposite. The couples were extraordinarily prolific, averaging 11.4 children. And these children grew strong and tall. The men of the first generation on Pitcairn were taller than their British fathers (and the average Tahitian man at that time) by two and one-half inches.

The mutiny on the *Bounty* is a dramatic story. The impressive number and stature of the half-British/half-Tahitian children born on Pitcairn is not surprising to me. It fits in perfectly with my theory. This first generation of Pitcairners no doubt received very different genes from their British fathers and Tahitian mothers. Their high heterozygosity allowed them to develop tall and (I assume) healthy, symmetrical bodies. This is still just anecdotal evidence, but it fits in perfectly with all of the scientific data I have presented.

A LITTLE BIT OF POP CULTURE

I prefer to focus on hard science, but anecdotal evidence is not hard to find. A list of multiracial models, actors, musicians, and athletes reads like a who's who of today's celebrities. And according to the conventional wisdom, biracial babies are cute, and biracial adults are exotic and attractive. Of course, conventional wisdom is often more conventional than wise. After all, in earlier generations the conventional wisdom told us that mixed individuals were inferior.

Today pop culture seems eager to embrace multicultural and multiracial trends, but in a strange throwback to the "one drop" rules of slavery, mixed individuals often aren't categorized as such. Halle Berry is a beautiful and talented actress who is generally thought of as African American, even though she is half white and half black. Berry seems to view herself that way too; in an emotional acceptance speech, she dedicated her Oscar to past and current black actresses and "every nameless, faceless woman of color that now has a chance because this door tonight has been opened."

Tiger Woods is an even more extreme case of this. At the tender age of twenty-one, his golf prowess stunned the world. Soon he was everywhere. You can now buy Tiger Woods clothing and golf clubs, as well as the Tiger Woods wristwatch and video game. You can read Tiger's book on how he plays golf, or you can read his dad's book on how he trained Tiger to play golf. In short, Tiger Woods has become something that was previously an oxymoron: a superstar golfer.

Because of his success on and off the links, Tiger quickly emerged as The Great Black Hope. But this was a title that he wasn't so eager to wear. Both of Tiger's parents are mixed and according to his description, he's only ¼ black. He pointed this out in a press conference and described himself as "Cablinasian" (a word he invented to reflect his Caucasian-Black-Indian-Asian heritage). According to *Time* magazine, this caused a backlash from many blacks who "saw 'Woods' assertion of a multiracial identity as a sellout," and looked at him as a "traitor."

It's not surprising that Woods was pretty reluctant to discuss race after the Cablinasian debacle. But he recently opened up in Charles Barkley's book about race in America titled, *Who's Afraid of a Large Black Man?* Woods writes, "I became aware of my racial identity on my first day of school, on my first day of kindergarten. A group of sixth graders tied me to a tree, spray-painted the word 'nigger' on me, and threw rocks at me. That was my first day of school. And the teacher really didn't do much of anything." Clearly, being pigeonholed as black was not a new experience for Mr. Woods.

The accomplishments of Tiger Woods, Halle Berry, and other interracial celebrities are impressive, but it's impossible to know how much of their success to attribute to their mixed heritage. Anecdotal evidence is fun, but it doesn't have the rigor of a carefully controlled scientific study. Still, before I return to discussing hard science, I'll indulge in one last bit of celebrity gossip.

ONLY JULIA KNOWS FOR SURE

I'm a big fan of Julia Roberts. Her talent and charisma have catapulted her to a successful career as the most highly paid actress of her generation. She is truly America's Sweetheart. Because of her celebrity, her life is under constant scrutiny. Her romantic affairs are of particular interest to her public, and she has given her fans plenty to talk about. Roberts has had many relationships, most with high-profile, handsome men.

One of those handsome men was Benjamin Bratt. Roberts and Bratt dated for three and one-half years starting in 1998. Bratt is a successful actor in his own right. He's also biracial—born to a German/English father and a Peruvian Quechua Indian mother.

In the mid-1990s, a few years before dating Benjamin Bratt, Julia was briefly married to Lyle Lovett. The most common reaction to the Roberts-Lovett coupling was surprise. The reason people were so surprised is simple: Mr. Lovett is not handsome. I feel bad saying it, but it's true. Lovett is a very talented singer-songwriter, but in the looks department he's not much to write home about. At no time was this more evident than when he was standing next to the beautiful and glamorous Julia.

As if it weren't bad enough that all of America was discussing his unattractiveness, we also have scientific confirmation of it! When *Newsweek* magazine ran a cover story on the science of beauty, Lyle Lovett was their poster boy for asymmetry. Most asymmetries are subtle, but not Lovett's; his lopsided features are obvious in pretty much any photo of him. *Newsweek* created a symmetrical Lyle by replacing the right half of his face with a mirror image of his left. As you'd expect, he looks a lot better in this version, though sadly, he's still not good looking enough to be with Julia Roberts.

So we have two men, both lucky enough to have been involved with Julia Roberts. We have Benjamin Bratt, whose good looks and biracial heritage have symmetry written all over them. And we have Lyle Lovett, who is a real life Lopsided Lou. And, as you recall from Chapter Two, we have a study showing that highly symmetrical men are more than twice as likely to bring their partner to orgasm during intercourse as low-symmetry men.

I feel a little guilty speculating about what went on in Julia Roberts's bedroom, but the implications are clear. In pretty much any scenario, I'm betting on Benjamin Bratt. Of course, I can speculate all I want. At the end of the day, only Julia knows for sure.

Milk and Turkeys

Some may find the topic of this book to be surprising or controversial, but farmers have been exploiting these same principles for generations. Humans began domesticating plants and animals thousands of years ago. They quickly learned that by selectively breeding certain plants or animals they could maximize desirable traits and minimize negative ones. With these methods a farmer can increase the yield of his crops with less or the same amount of work, often with very impressive results.

In the early part of the twentieth century, these methods were applied to dairy cows. In 1909, the average American drank 34 gallons of milk per year. Milk consumption reached an all time high of 45 gallons per person per year during World War II, when many other foods were being rationed. Today, we are each drinking less milk—closer to 23 gallons of milk per year—but eating 30 pounds of cheese, 8 times as much as 100 years ago. Bottom line: Americans love their dairy.

If you're a dairy farmer, you want to produce as much milk as possible. More milk from fewer cows means more profits. An aggressive campaign of selective breeding accomplished just that. From the 1920s to 1950s, just by selectively breeding the cows that gave the most milk, farmers were able to more than double average milk production. A cow can now produce over 50 pounds of milk per day!

If there's anything more American than a nice glass of milk, it's a turkey dinner. It's the centerpiece of Thanksgiving and the cornerstone of festive dinners and lunchbox sandwiches everywhere. As a matter of fact, the turkey is such a part of American culture that Benjamin Franklin lobbied for it to be the national bird! Happily, cooler heads prevailed.

Just as dairy farmers wanted more milk, turkey farmers wanted more meat. Specifically white breast meat, which Americans favor over dark meat. Enter selective breeding. In a feat that would impress even a Beverly Hills plastic surgeon, farmers have been able to significantly increase the size of the average turkey breast. Not with silicone or saline, but just by breeding the turkeys with the most breast meat. In this case the plan worked almost too well. Male turkeys now have such large breasts that they cannot successfully mount females. Their giant chests get in the way! This renders them completely unable to mate and, as a result, farmers have had to step in. Every Thanksgiving drumstick, turkey sandwich, or turkey pot pie that you eat comes from a turkey that was created with artificial insemination. Selective breeding has given us the big-breasted turkeys we need to satisfy our national hunger for white meat and, in the process, completely eliminated turkey sex.

You only have to look at a Chihuahua sitting next to a Great Dane to recognize that selective breeding can lead to some pretty dramatic results, but what does that have to do with interracial people? When farmers select for a trait, they mate two individuals who both possess that trait. The turkeys with the biggest breasts, for example. So selective breeding involves mating individuals that are very similar. It's the opposite of everything we've been talking about.

Isn't that bad? Well, yes and no. It's good in that you can emphasize a certain trait, as we've seen with dairy cows and turkeys. It's bad in all the ways we've discussed: mating similar individuals leads to loss of genetic diversity and poor symmetry, attractiveness, and health. Farmers know about this downside to selective breeding and refer to it as inbreeding depression. Mates that are similar often have inferior offspring in terms of health, fertility, and yield.

Farmers are very familiar with inbreeding depression, but are often willing to live with the negative consequences in order to select for a particular trait, like big turkey breasts. However, because farmers know about inbreeding depression, they also know about its opposite.

HYBRID VIGOR

Selective breeding is all about creating separate, somewhat inbred lines of a crop or animal. At some point, farmers discovered that crossing these very different lines could have surprising results. The first generation born of these highly varied parents often possesses amazing attributes. This hybrid generation can greatly outperform both parents in terms of hardiness, growth, and output. Farmers refer to this boost as hybrid vigor.

This book is about hybrid vigor. Some may find it scandalous to apply it to people, but farmers have known about and applied this technique for generations. Crossing two separate strains of wheat or apple or chicken produces offspring with high levels of genetic variation. We know how powerful high heterozygosity can be. So do farmers. The average farmer may not know the word "heterozygosity," but the benefits of hybrid crosses are obvious.

Much of the early work on hybrid vigor in agriculture was done on corn. There are two important traits that make corn ideal for hybrid crosses: 1) corn is wind pollinated, and 2) the male part (the tassel) and the female part (the ear) are pretty far apart on a corn plant. If you are interested in growing corn, you surely want to take advantage of the increased yield caused by hybrid vigor. Here's how to grow your own hybrid corn at home, in five easy steps:

1. Get two different strains of inbred corn. We'll call them Corn A and Corn B.
2. Plant them in alternating rows. Corn A, Corn B, Corn A, Corn B, and so on.
3. Cut the tassels off of all of the Corn B stalks. (Apparently this is a popular summer job for teenagers in corn producing areas.)
4. Wait. The wind will carry pollen from the Corn A tassels and fertilize the Corn B ears. Since the Corn B rows don't have any tassels, they won't produce any pollen, and you don't have to worry about self-fertilization. All of the Corn B ears will be fertilized by Corn A pollen and will thus produce hybrid corn.
5. Plant the hybrid seed from the Corn B rows.

Then, with some water, sunshine, and patience, you get hardy, high-yield, hybrid corn.

The disadvantage of hybridization is that you have to re-create the hybrid seed every generation. To maximize the genetic diversity of the offspring, you have to start with two very different parents. Sound familiar? The same is true for people. Despite the inconvenience of re-creating hybrid seed every generation, the significant advantages of hybrid vigor make it worthwhile for farmers.

Early research on corn hybrid vigor was done by George Shull around 1907. His work paved the way for large-scale application of hybrid vigor for corn and other crops. In 1933, hybrid corn only accounted for about 1 percent of the total crop. Ten years later it was 50 percent. By the 1960s, almost all corn in the United States was grown from hybrid seed. The widespread adoption of mixed seed is not surprising considering that hybrid corn yield can be three times that of purebred lines. This has helped corn to become the most popular grain in the world and means more corn on the cob, corn chowder, corn tortillas, popcorn, grits, cornbread, and corn flakes for all of us. Traditionally, farmers predicted that if the corn was "knee-high by the Fourth of July" it would be a good harvest. This is far more likely with hybrid seed.

Other crops have also benefited from hybridization. Wheat, cotton, and alfalfa have made significant gains as a result of increased genetic variation. And hybrid vigor isn't just relevant to plants. Research at Texas A&M University has shown that mixing different breeds of cattle results in a 10 to 20 percent calf crop increase. That means more burger for their buck.

In addition to the general benefits of hybrid vigor, crossbreeding can sometimes mix desirable traits from the two parent strains. Historically, wheat yield was held back because the stalk could only support a certain amount of grain. If the kernels were too heavy, the stalk would buckle under their weight. In the 1950s, this limitation was overcome via hybridization. Crossing dwarf wheat, which has shorter, sturdier stalks, yielded two important benefits. Just as with corn, the hybrids were vigorous and prolific, but they also inherited the dwarf stems that allowed them to support heavier kernels. Thus, they wouldn't be crushed under the weight of their own growth. Similar success has been achieved with rice stalks.

Rin Tin Tin's Granddad

The creation of purebred dogs is a classic example of selective breeding. Each breed was created to maximize certain desirable traits in appearance or behavior. Developing a breed and continuing its line both involve inbreeding, especially if the available breeding pool is small. German Shepherds are one of the most popular of the purebreds. I was surprised to learn that the breed is only about one hundred years old. For many generations, native sheepdog breeds were popular in Europe, but were not rigidly organized and classified like they are today until Captain Max von Stephanitz appeared on the scene.

In 1899, von Stephanitz saw a dog that he felt exemplified the best traits of a sheepdog. He purchased this dog, renamed him Horand v Grafeth (maybe "Spot" was already taken), and formed the Verein fur deutsche Schaferhunde (or SV). The SV was a society designed to promote this new breed, a breed started with a single individual! If every German Shepherd alive today is a descendent of Horand v Grafeth, there has clearly been considerable inbreeding. This has allowed breeders to carefully control the hallmarks of the German Shepherd, but also means a loss of genetic diversity and sometimes the inclusion of undesirable genetic traits. German Shepherds are vulnerable to a whole host of genetic disorders, including hip dysplasia, degenerative myelopathy, and intervertebral disc disease.

In general, purebred dogs are created through programs with significant inbreeding and suffer from many genetic problems. *The Control of Canine Genetic Disease*, a book that covers 380 dog breeds, lists 532 genetic disorders. The German Shepherd is predisposed to 132 of them! The only breed affected by more genetic disorders is the Poodle.

According to the Poodle Club of America, Poodles are at risk for hip dysplasia, progressive retinal atrophy, cataracts, seizure disorders, thyroid disorders, Cushings, and von Willebrand's disease, among others. It's easy to understand the conventional wisdom that mutts are healthier and longer lived.

I'm happy to report that recently there has been a new interest in mixed breed dogs, particularly mixes involving Poodles. Just like dwarf wheat, Poodles possess a very desirable trait that breeders would like to incorporate into future generations: Poodles are hypoallergenic. Well, sort of.

Poodles are popular among allergy sufferers because they shed very little hair and dander. Many breeders have attempted to crossbreed Poodles in the hopes of maintaining this desirable feature in a new type of dog. The most famous of these is the Cockapoo, a Cocker Spaniel/Poodle mix. A newer mix, with an equally silly name, is the Labradoodle, a Labrador Retriever/Poodle mix. The appearance of the Labradoodle varies, but is often that of a Labrador with a bad perm.

Poodles have also been crossed with Golden Retrievers (Goldendoodle), Pekingese (Pekeapoo), and Schnauzers (Schnoodle). I'm not making these up. While the offspring of such unions do not necessarily inherit the Poodle's hypoallergenic coat, they certainly inherit a healthy dose of hybrid vigor. Sadly, it's hard to teach an old dog breeder new tricks. Many breeders are attempting to turn these crosses into separate breeds that will be recognized by kennel clubs. This usually involves a systematic program of inbreeding.

Although hybrid corn and Labradoodles don't appear to have a lot in common, their genetic history is comparable. All our examples of hybrid vigor follow a similar pattern. Two or more strains or variants are created by inbreeding a small number of individuals. These strains are kept separate and only breed among themselves. Over time, genetic diversity is lost, and the individuals in this strain become more homogeneous. As different versions of genes are lost, there is less and less heterozygosity.

Generally, each strain will lose some versions of genes and keep others. Since the different strains will keep different versions, they will become genetically distinct. They are still part of the same species, but are recognizably different. When two of the strains are crossed, these different gene versions come together, and the progeny have high levels of heterozygosity. These offspring have the genetic variation required to cope with a varying environment and develop according to their DNA blueprints, with impressive growth, vitality, and fertility. Voila! Hybrid vigor.

The parallels to human history are obvious. Agricultural hybrid vigor is usually the result of an intentional breeding program. In humans, it's just a consequence of geographic and social isolation. For most of our history, we lived in small groups, and there was little mixing. Small group size (often compounded by population bottlenecks) and some inevitable inbreeding led to the development of genetic and physical differences. This was basically the creation of human strains. We call these strains, "races." Mixing two different inbred strains leads to hybrid vigor whether you're talking about wheat, dogs, or humans. It's as simple as that.

WHAT THE HELL IS A ZONKEY?

In Chapter One, I discussed the prevailing scientific view in the nineteenth century that mixed race individuals were physically and mentally inferior. The opposite of hybrid vigor. The use of the term "mulatto" reflects this view. This word most likely comes from the Spanish word, "mula" which means mule. Half horse and half donkey, the mule is a mixed animal. All mules have a horse mother and a donkey father. (A donkey mother and a horse father produce a mulelike animal called a hinny.) Mules don't have mule parents because they are sterile. Every single one of them. Since mules are mixed and flawed, mulatto seemed like an apt name for biracial humans back in the day. In Chapter One, I promised, "In Chapter Five, we'll see why mulatto is a misnomer." Well, here we are. So why is mulatto a misnomer? And why don't mules have hybrid vigor?

Maybe mules do have some hybrid vigor. According to the American Donkey and Mule Society, mules have more stamina than horses and can carry more weight. Still, complete infertility is a pretty big problem. What went wrong? The important distinction is that all of the mixing we've discussed so far is between different strains (or races) of the same species. Horses and donkeys are not different strains; they are two completely different species.

Horses and donkeys look pretty different, but even more important they are very different on a genetic level. Specifically, they have a different number of chromosomes. Your DNA contains all of the genes that build and run your body. These genes are stored on long strands of DNA called chromosomes. Every human should have the same number of chromosomes: 46. These chromosomes are divided into 23 pairs. You inherited 23 chromosomes from your mother and 23 corresponding chromosomes from your father. These 46 chromosomes met up and formed the full complement of your DNA in the fertilized egg that later developed into you.

Horses have 64 chromosomes while donkeys have only 62. Normally each chromosome has a corresponding twin. This is impossible with a mule. Mismatched chromosomes are a serious problem, and as a result, most interspecies love affairs don't lead to any offspring. Horses and donkeys are similar enough that they can mate and produce hardy, viable babies, but the chromosome mismatch means they won't be getting any grandchildren.

While interspecies breeding is rare, it does happen. Donkeys can also mate with zebras, resulting in zonkeys. A zebra/horse cross is called a zorse, or zebroid. And, depending on which is the father and which is the mother, a lion and tiger can produce ligers and tigons. (Fans of hipster antihero Napoleon Dynamite will remember that the liger is "pretty much" his favorite animal.)

Sheep and goats, while similar looking, have an even greater chromosome mismatch. Sheep have 54 chromosomes, while goats have 60. Because of this, sheep/goat hybrids are almost always stillborn. However, there is one documented case of a viable sheep/goat hybrid. Born at the Botswana Ministry of Agriculture, this unique creature had a goat mother and a sheep (ram) father. Like mules, the sheep/goat mix appeared to be infertile, but may have shown some signs of hybrid vigor—he grew at a faster rate than other kids and lambs born the same month. In fact, he was overly vigorous. His tendency to mount both sheep and goats, even when they weren't in heat, earned him the nickname "Bemya" or rapist. Eventually this unusual hybrid had to be castrated to reign in his amorous tendencies.

Interspecies crosses are weirdly intriguing, but rarely produce healthy offspring. In general, I'd advise you to stick to your own species.

BACK TO THE WILD

Inbreeding depression and hybrid vigor are not just issues for domesticated plants and animals. In the wild, inbreeding and outbreeding can have significant effects.

In the early 1970s, seventeen dama gazelles and seventy-two dorcas gazelles were transplanted from the Western Sahara to the

Estación Experimental de Zonas Aridas in Almeria, Spain. Since then, the gazelles have done the things that gazelles do: ran around, foraged, had baby gazelles, and learned Spanish. (OK, maybe not that last one.) Since the scientists there keep a close eye on things, they have detailed life histories for each of the animals, including how they are related to each other. This coupled with the small group size makes for an ideal situation to study inbreeding.

When researchers measured the gazelles' skulls for nine different measures, they found that some were more symmetrical than others. No surprises there. When they compared these data to the gazelles' family trees, they found that the more inbred individuals were the least symmetrical.

The transplantation of a small number of gazelles to Spain is an example of a population bottleneck. When a population dramatically decreases in size, a lot of genetic diversity is lost. This can happen because a small subset moves to a new place, as with our Spanish gazelles (this is called Founder Effect), or because a majority of the population dies leaving only a few survivors. Either way, the members of the new, small population probably don't contain all the gene variants that the original population did. Plus, they have fewer options for mates, which can lead to inbreeding.

It's rare to be able to study a group before and after a population bottleneck, but scientists have been able to do just that with the northern elephant seal. These seals got their name because the males have a trunklike nose that they use to produce ear-splitting roars during mating season. And because they are *huge*. The largest elephant seal on record is a twenty-two foot long male that weighed in at 7,500 pounds. That's about the size of two Toyota Corollas! A lot of that weight is blubber, a layer of fat that provides energy storage and important insulation in the chilly ocean waters. This blubber is also the reason for the elephant seal's population bottleneck.

In the nineteenth century, marine blubber was a valuable commodity that was rendered down into oil. The elephant seal was hunted for its blubber, almost to extinction. By the 1890s, only ten to thirty seals remained. Happily, since then the population has made an astounding recovery and now numbers well over one-hundred thousand seals. A classic population bottleneck.

Scientists were able to compare skulls of pre–bottleneck seals collected in the nineteenth century (stored in the Smithsonian Institution) to modern-day, post–bottleneck seal skulls. The skulls were measured for symmetry, and DNA was extracted for genetic analysis. As we would expect in a bottleneck situation, the researchers found a loss of genetic diversity in the post–bottleneck seal samples. They were also less symmetrical than their pre–bottleneck seal ancestors.

Human history is filled with population bottlenecks caused by disaster or a small group splitting off to form a new population. Some DNA evidence points to a severe human population bottleneck around seventy-thousand years ago, possibly caused by the eruption of a supervolcano in Indonesia. Other bottlenecks only affected specific groups, such as when a small group of Vikings colonized Iceland about one-thousand years ago. In any case the result is the same—a loss of genetic diversity and all of the negative consequences that go along with that loss. Only one group of northern elephant seals survived their population bottleneck, so there is no way for the seals to recover their lost genetic variation. We humans are luckier. We have different genes stored in different populations, and some of our diversity can be restored just by mixing, with all of the accompanying benefits.

Laboratory experiments have shown this time and time again. When inbred strains of house mice were mated, their hybrid babies were significantly more symmetrical at 1 month, 3 months, and 5 months of age. A similar experiment showed that hybrid water fleas had greater survival rates than their inbred peers.

Some organisms can choose between inbreeding and outbreeding. Like many plants, the cucumber can self-pollinate or be cross-pollinated with the pollen from another plant. Because of everything you've read thus far, you know that cross-pollinating will produce healthier cucumbers. Surprisingly, the cucumber plant knows that too. If the plant has both self- and cross-pollinated cucumbers, it will selectively abort the self-pollinated ones, which are highly inbred. This way the plant can focus its resources on the more viable, heterozygous, cross-pollinated fruit. Charles Darwin wouldn't be surprised to hear this; he once wrote about orchids, "Nature thus tells us, in the most emphatic manner, that she abhors perpetual self-fertilization."

This is good stuff to know if you're a mouse or cucumber, but what about people? All these animal studies provide parallels for human populations, but there's no substitute for actually looking at good old *Homo sapiens*.

HUMAN HYBRID VIGOR

As I mentioned in the last chapter, most scientists are not eager to delve into race. The controversial nature of the topic means that studies of this type are rare, but some do exist. A 1957 study examined the lives of villagers in a region of Switzerland called Ticino. This beautiful area is the only state in Switzerland located south of the Alps, and surprisingly its official language is Italian. Although many of these Swiss villages are located close together, they are tightly knit communities, and historically most couples both came from the same village. The author of the study compared those whose parents came from the same village with those whose parents came from two different villages. He found that the men who were a blend of different villages were taller, bigger in the chest, and heavier.

This study isn't obviously about interracial unions. Most people would consider all of the Swiss villagers to be of the same race. But remember, race is a continuum. Because of generations of intermarriage, the citizens of Village A are more similar to each other than they are to the citizens of Village B. Thus, the children of a Village A/ Village B marriage have some genetic advantages. If a member of Village A married someone from outside of Switzerland, those advantages would be even greater. Even more so, if said villager married a non-European.

A similar study was carried out in Poland in 1970. Here, citizens of the Polish town of Szczecin were examined, and their life histories recorded. The researchers focused on the birthplaces of each person's parents. In theory, the farther apart their birthplaces, the more genetically different they are, and the more hybrid vigor we should see in their children. And that's exactly what they found. I don't know how to pronounce Szczecin, but I do know that as the distance between the parents' birthplaces increased so did the height, weight, and chest circumference of their children.

Just as with our Swiss friends, here Poles were marrying Poles, but not all Poles are created equal. And not all Poles are equally related to each other. Even within a nationality or race, there is a lot of variation. Of course, interracial mixes should provide even more heterozygosity and a correspondingly bigger dose of hybrid vigor. So let's take a look at some interracial studies.

EAST MARRIES WEST

Hawaii has always been a meeting point between East and West. This tropical paradise has no ethnic majority; its two biggest minorities are Asian Americans and European Americans. Possibly because of its tremendous ethnic diversity (or because of the abundance of piña-coladas), Hawaii has the highest rate of intermarriage of any state in the union. According to the 2000 Census, an incredible 21.4 percent of Hawaiians identify themselves as belonging to two or more ethnic groups. Hawaii is a true melting pot.

When scientists did a study in Hawaii on the genetics of intelligence, they were guaranteed to run into some interracial individuals. They focused their research on Americans of European Ancestry (AEA) and Americans of Japanese Ancestry (AJA), but some of their participants were both. The researchers administered a battery of 15 cognitive tests designed to measure intelligence. All of the groups had a similar socioeconomic background, but the biracial, half-European/half-Japanese participants outscored both the AEA group and the AJA group on 13 of the 15 tests.

You Really Are
Smarter Than Your Parents

The cognitive performance of the biracial group is impressive, and it may hold the key to a long-standing mystery. For some reason, we seem to be getting smarter. No one is sure why, but the average IQ score has consistently been creeping up by about 3 points every ten years. This implies that each generation is significantly smarter than the previous one.

The phenomenon, known as the Flynn Effect, after its discoverer, has been seen all over the world. The effect is commonly believed to be caused by some kind of environmental factor. The most popular theory is better nutrition. Remember how your mom always told you to eat a good breakfast before a big test? Even though she came from a dumber generation, she was right. Better nutrition does lead to improved physical and mental development. But at the same time, nutrition can only go so far. And the link between nutrition and the Flynn Effect is somewhat shaky. The effect was stronger than ever in Scandinavia in the years following World War II, even though during the Nazi occupation, food was almost as scarce as freedom.

At least one intelligence expert thinks that the cause isn't environmental at all—it's hybrid vigor. True interracial marriage is still relatively rare, but the world is a smaller place than it's ever been. People are less and less likely to marry someone from their own village. Plus, increased immigration and the growth of large cities mean that even if you marry the boy or girl next door, they may be ethnically very different from you. So, even if people are marrying within their race, they are still probably producing children with higher heterozysogity than the previous generation. That higher level of heterozygosity may account for the mysterious IQ boost.

The average height and growth rate have also increased significantly over the past few generations. Just as with intelligence, better nutrition usually gets all of the credit. Certainly, nutrition is a factor though I don't think it's the whole story. It wasn't long ago that diseases like scurvy and beriberi were significant public health problems. Scurvy is caused by a lack of vitamin C and sufferers complain of weakness, bruising, and bleeding gums. A diet without fresh fruit or greens can lead to scurvy. Beriberi is caused by a deficiency in vitamin B_1 (thiamine). People with beriberi suffer from extreme weakness and heart problems. It can be fatal. The name beriberi comes from the Sinhalese language (spoken in Sri Lanka) and means, "I cannot. I cannot."

Nutritional deficiencies like these are still a problem in some parts of the world, but they have mostly been conquered in first world countries. When was the last time you called in sick because of a bad case of scurvy? Certainly nutrition has significantly improved over the last few hundred years. I have no doubt that it is a big part of the gains in intelligence, height, and growth rate that have been recorded. But I also believe that hybrid vigor is another important component of these improvements.

As the world has become a smaller place, people have more choices of marriage partners. Most of us are no longer limited to the population of our village. Even though a majority of us are still marrying people from the same ethnic group, we are mixing genes more than ever before. When my friend Lorna, whose grandparents were Italian immigrants, married Stefan, who is of Swedish extraction, it was hardly scandalous. To most people they both just seem "white." But this is a pairing that was geographically and socially unlikely in previous generations. They are more genetically distinct than an Italian-Italian or a Swedish-Swedish couple would be, and their kids will benefit from increased heterozygosity.

I see this as the first phase of human hybrid vigor. And I think it's at least partially responsible for the Flynn Effect, as well as other modern-day improvements. By introducing even more genetic variation, we can go even further. Once true interracial marriage becomes commonplace, I think we'll see an even bigger boost in the physical and mental development of future generations. I find this thought incredibly exciting. I like the idea that with everything that we've achieved as a species, we can still do better. We can still do more. We have enormous potential lying untapped in our DNA, just waiting to be released. It's also rather poetic that after every inappropriate racist joke, hurtful cultural slur, destructive race riot, and horrible ethnic cleansing, the key to so much lies in embracing our differences. It's ironic and sad that wars are still being fought over ethnic divides, when everything you could ever wish for your children—health, beauty, intelligence—is buried inside the cells of your enemy.

Frankenstein vs. Stalin?

The small state of Meghalaya in Northeast India is known for heavy rains and a weird sense of humor. All three of Meghalaya's major tribes have Laugh Clubs and apparently even naming your children is an opportunity for a joke. Recent elections in Meghalaya featured politicians named Frankenstein W. Momin, Stalin L. Nangmin, and Tony Curtis Lyngdoh, among others.

The largest of Meghalaya's tribes is the Khasi. While some Khasi still practice the traditional religion of Niam Khasi, about two-thirds are Christian, thanks to the efforts of Presbyterian missionaries in the nineteenth century. The Khasi have contact with other groups, but intermarriage is rare (possibly because foreigners are worried about bringing Frankenstein home to meet the folks).

Sometimes the Khasi do intermarry with other groups, including Muslims that have migrated to the area from Bangladesh and other regions of India. It is far more common for a Khasi female to marry a Muslim male, and a recent study looked at the children of this type of couple. According to the author of the study, "From the anthropological point of view, Khasis and Muslims of Northeast India belong to the Mongoloid and Caucasoid racial groups, respectively." Mongoloid and Caucasoid are broad racial categories that have fallen out of favor, and some people even find offensive. Regardless of labels, however, the Khasis and Muslims come from very different ancestries and should produce children with high levels of heterozygosity. This study focused on the daughters of such marriages and found the hybrid vigor that we would predict. Khasi/Muslim girls consistently beat out Khasi/Khasi girls in height, sitting height, and growth rate.

I ♥ Symmetry

I love symmetry. I really do. I find it fascinating that it gives us such powerful information about what's going on in another person's genes and that it's linked to so many important traits. It's incredible that it can predict so much about a person's attractiveness, health, intelligence, and athleticism. You shouldn't be surprised that I like symmetry so much; I did write a whole chapter about it. So to me, the real currency of hybrid vigor is symmetry. Interracial people have greater genetic diversity which means they are more symmetrical. It also means that they are smarter, taller, and grow faster, but the real crux is symmetry. We know that symmetry is the goal of your developing body. If you've achieved it, your body's done everything right. If you've got symmetry, you've got it made. Let's take a look at interracial people and symmetry.

Recently, Dr. Jay Phelan examined the question of interracial symmetry in a groundbreaking study at UCLA. Phelan recruited undergraduates who were either biracial (parents are two different races) or uniracial (parents are the same race). He used five broad racial categories: black/African American, white/European/Middle Eastern, American Indian/Alaskan Native, Asian/Pacific Islander, and Hispanic/Latino. Both the biracial and uniracial participants were measured for symmetry. They each also had their photo taken. Later another group rated the photos for attractiveness.

It's very easy for me to tell you that the study found the biracial participants to be more symmetrical and attractive than their uniracial peers. In fact, that's what happened. But that really doesn't capture the dramatic findings of this study. The truth is that the biracial group blew the uniracial group out of the water.

Often in science, results are subtle. If one group is ten centimeters taller than another it's easy to call that a real difference. But what if the group is only one centimeter taller? Or one-quarter of a centimeter? How do scientists differentiate between a real effect and just a small random difference? We use statistics. With statistics, we can calculate the magnitude of the difference and the odds of it being random chance. It all comes down to a magic number called a p value. p is the likelihood that the effect is just due to chance. The stronger your effect, the lower p is. The gold standard of scientific studies is $p = .05$. This means there's only a 5 percent chance that we're seeing random fluctuations and a 95 percent chance that we're seeing a real effect. If you have a p value of .05 or lower, you know you're seeing a real result.

When Dr. Phelan compared the symmetry of the biracial group with the uniracial group, he got a p value of .00007! That means there's only a .007 percent chance that this is random noise and a 99.993 percent chance that biracial people really are more symmetrical than uniracial people. In a world where .05 is good enough, that's huge. A quick look at the data is just as obvious. Of the 25 least symmetrical participants, how many were biracial and how many were uniracial? Every single one of the bottom 25 was from one of the uniracial groups. What about the top? Of the 8 most symmetrical participants, 7 were biracial. This is not a subtle finding. The biracial group was dramatically more symmetrical.

If you remember only one study from this book, remember this one. Actually, who am I kidding? You're going to remember the one about women orgasming with symmetrical men. But try to remember this one too. After all, they're related—just ask Benjamin Bratt.

A REALLY GOOD QUESTION

I've already told you about the two questions that I always get when I give talks on interracial advantages. But on rare occasions I get another, more interesting question: If interracial people really have these advantages why do so few people marry outside their race? Why haven't we evolved to seek out mates that are different from us? I love when I get this question because it displays a good grasp of evolution. In general, evolution moves us toward the optimal scenario. So what's going on here? I'm saying that the optimal situation is interracial marriage, but we know that it's a rarity. Has evolution let us down? To answer this question we have to delve into a complicated but fascinating topic: mate choice—who we find attractive and why. I'll tell you about the biological and social aspects of mate choice in the next chapter.

6

Why Do Fools Fall in Love?

BIOLOGY VS. SOCIETY IN MARRIAGE

IN JANUARY 1936, King George V of England died. What started as an unremarkable passing of the throne turned into what H. L. Mencken called, "the greatest story since the crucifixion." George's firstborn son, the new King Edward VIII, was expected to find a suitable queen to rule by his side. Although technically single, Edward was in a covert relationship with Bessie Wallis Simpson. He would have been hard pressed to find a worse candidate to be his bride.

Wallis Simpson was American, divorced, and still married to her second husband. Even one divorce was unforgivable in these conservative times, but if she were to marry Edward she would be twice divorced! And to make matters worse, she was Catholic. The king is also the head of the Church of England, and the British constitution specifically forbids him from marrying a Catholic.

Parliament gave the young king an ultimatum: give up Wallis Simpson or give up the throne. On December 12, 1936, King Edward VIII abdicated in favor of his younger brother. He told his people, "I have found it impossible to carry the heavy burden of responsibility, and to discharge my duties as king as I would wish to do, without the help and support of the woman I love." Edward and Wallis Simpson were married a few months later.

Depending on how cynical you are, you may see Edward's choice as either incredibly romantic or incredibly foolish. The more idealistic among you will be happy to hear that the couple stayed married for 35 years, until Edward's death in 1972. It's impressive that Edward gave up everything for his lady love, but he did want to be king. He negotiated with Parliament in order to have both Wallis Simpson and the monarchy, but failed. Why didn't he just pick a more suitable girlfriend? Why did he fall in love with such an inconvenient woman?

More Than Skin Deep

The choices that we make in love are complex and are influenced by a myriad of biological and social factors. In many cases they're not even choices, we're just following the rules of attraction encoded in our genes. Critics often decry any attempt to look at the biological basis of beauty and attraction. They claim that the standards of good looks are set by the media and that beauty is just a social construct. Much like those who claim that race is merely a social construct, these naysayers are wrong.

Certainly some aspects of beauty are cultural. Fashions and trends change over time. I look back at photos of myself from the eighties and can only shake my head in confusion. My DNA wasn't telling me to wear pastels and grow a mullet—society was. And for some bizarre reason, I listened. Similarly, some aspects of beauty vary widely across cultures. Giant plates inserted into holes cut in one's upper or lower lip are the height of fashion among the Botocudo of South America. Oddly, that just doesn't do it for me.

Many standards of beauty, however, are universal across time and culture. People in different parts of the world consistently find photographs of the same people attractive. One study had British, Chinese, and Indian women rate photos of Greek men. Their responses were all the same. A handsome man is a handsome man, whatever your culture.

To quell critics who say we are all just responding to the western media, some researchers tried similar studies on the Ache and Hiwi tribes. These groups live in Paraguay and Venezuela, respectively, and, for the most part, live a traditional lifestyle of isolated foragers. In any case, they don't subscribe to *Vogue* or *Cosmo*. But they found the same people attractive as the European raters.

A similar result was found by a psychologist at the University of Texas. She showed photos of faces that had been rated either attractive or unattractive to three and six month old babies. It didn't matter if the faces were of white females, white males, black females, or even other babies. The little guys stared at the attractive faces significantly longer. These babies don't watch MTV, but they find the same faces attractive as adults. They are responding to built-in rules about beauty.

So What Are the Rules?

Remember, evolution boils down to one thing: making babies. Seeking out food and other resources, competing for mates, avoiding predators—they're all just part of one big baby-making contest. Attractiveness and beauty are just concepts that help us choose good mates. The innate rules of sexiness help us find the best possible person with whom to make babies. Of course, when I meet a girl, I'm not consciously evaluating her as a potential mother for my children. Motherhood is the last thing on my mind, but on a physical level, my body responds to traits that make her a good candidate.

We've already talked about one of the most important of these traits: symmetry. It's sexy because it's an honest sign that this person has a good set of genes that have been able to construct a body according to plan. But symmetry is not the only indicator we look for. Anything that gives us information about the other person's health and fertility is useful. And the more health and fertility we see, the more turned on we get.

Clear skin is a perfect example. As much as fashion and beauty can vary across cultures, clear skin is universally admired. Our society has seen some strange trends come and go, but there's never been a wave of Leprosy Chic. Why? Because clear skin means a clean bill of health: no parasites or disease here. I may not know what the mother of my children will look like, but when I imagine her I don't see skin lesions and ringworm. Our love for clear skin steers away from diseased individuals and toward healthy, fertile potential mates.

Like Sand through the Hourglass

In the old Three Stooges shorts, whenever one of the guys would see a hot "dame" he would whistle and make an exaggerated hourglass motion with his hands. Watching these slapstick antics as a kid, I was too young to fully appreciate a woman's curves, but Larry, Moe, and Curly certainly did. It turns out, they're not alone.

The appeal of a woman's hourglass figure is nearly universal. Curves separate the girls from the boys, and apparently they also separate the hot from the not. For centuries, women have known that a narrow waist and wide hips are appealing and have gone to great lengths to accentuate these features with corsets, bustles, and petticoats. Now scientists have calculated the optimal ratio of waist to hips. For maximum sexiness, a woman's waist should be 70 percent the size of her hips.

Different body types may go in and out of style, but this waist-to-hip ratio of .7 is constant. One study looked at Miss America winners from 1923 to 1990. The contestants in the early years were markedly plumper than they are today, but the waist-to-hip ratio of the winners hardly varies—every one was between .68 and .72. The same was true for *Playboy* centerfolds over the years. Even Twiggy, the it-girl of swinging sixties London, whose nickname is a reference to her stick-thin figure, was more hourglass shaped than you'd think. Her waist-to-hip ratio was .73.

Why are men so drawn to a .7 body type? Remember, it all boils down to babies. Women with a .7 waist-to-hip ratio are more fertile than women with other body types. One study found that every 10 percent increase in a woman's waist-to-hip ratio decreased her chance of conceiving by 30 percent! No wonder .7 is the magic number.

Other attractive female traits are also linked to fertility. Full lips, a small jaw, and delicate chin are all sexy features. They're also all linked to high levels of estrogen and thus indirectly to fertility. But don't worry; women aren't the only ones being judged this way. Men are rated more attractive when they have a strong jaw and thick brow—both caused by high levels of testosterone. This may explain why rugged-jawed men lose their virginity earlier and attain higher military rank than their peers.

EVOLUTION KNOWS BEST

In each of these cases, the rules of attraction are guiding us to a vital, fertile, desirable mate. One that will maximize the number of healthy, viable children we can have. Of course, all of the people who were attracted to diseased and infertile mates didn't do so well—they died out. We are lucky enough to have inherited desires that help us make better choices.

I've spent the last five chapters outlining the real and significant advantages of interracial mixing. If physical attraction is supposed to help us find the ideal mate, shouldn't it be steering us to marry someone of another race? Our subconscious impulses should be working overtime to ensure genetic variation for our kids. But do they? What are the rules of attraction when it comes to mixing?

OH BROTHER

In the spring of my seventh-grade year, I fell in love. Her name was Leslie. Ah Leslie. Even now, thinking about her, I can feel my heart swell. Her hair was like a river of honey and her eyes—I can't actually remember what color they were, but I'm sure they were beautiful. I had just started to appreciate curves and she had just started to grow some—it was a perfect match. I couldn't get her out of my mind. I'd lie awake at night, trying to think of words that rhymed with "Leslie."

Leslie and I started spending a lot of time together. I was ecstatic. We ate lunch together, talked on the phone, shared homework. Those were good times. And then one momentous day, Leslie told me that she loved me. Like a brother.

Needless to say, I was crushed. This was the greatest disaster I had faced in my twelve years of life. I didn't understand where I had gone wrong. Wasn't I looking good with my pastel shirts and my mullet? Certainly it was sweet that she loved me, but even at the tender age of twelve I knew the rules. And the rules say that a brother doesn't get to run his fingers through Leslie's gorgeous honey-blond hair. That would just be creepy.

And it would be creepy in any culture. Though different ethnic groups have very different mating practices, close incest is universally avoided. Marriages between a parent and a child or between two siblings lead to babies with incredibly high levels of homozygosity and all sorts of health problems. Just writing about it gives me a sick feeling in the pit of my stomach! But that sick feeling is important. It prevents us from making some really bad mate choices. We identify our family members very early in life—during the first two years. Once someone has been categorized as family, it's very difficult to see them any other way.

In the early days of Israel, many people lived on small collective farming communities called kibbutzim. These were communal societies and all aspects of kibbutz life were shared. Even child rearing. Children were not raised by their own parents, but communally in a common children's house. Since all of the children were raised together, they viewed each other as siblings. As adults, they just weren't attracted to each other and had to leave the kibbutz to find a mate. In almost three thousand marriages of people raised on a kibbutz, no one married a person they had been raised with. No one! Even though they weren't genetically related, their incest avoidance systems were triggered and couldn't be untriggered. They just couldn't see each other in a romantic light.

This phenomenon, called the Westermarck effect, after the anthropologist who discovered it, caused even worse problems in Taiwan. It's a desperate move, but sometimes poor Taiwanese couples sold their baby daughters to wealthy families. The rich family was happy to get some free labor and, once she had grown up, a wife for one of their sons. This practice was called *Shim-pua* which means "little daughter-in-law." You can imagine that this was a bad strategy; it has Westermarck written all over it. Since the *Shim-pua* and her future husband were raised together, being married felt creepy. *Shim-pua* couples were three times more likely to get divorced and were more likely to experience infertility and infidelity.

The Israelis and Taiwanese have seen the error of their ways. Communal child rearing and *Shim-pua* marriages have both fallen out of favor.

Here is a case where our biology is encouraging genetic diversity. Sometimes it gets confused, like with the kibbutz kids and the *Shim-pua* couples (and me and Leslie), but for the most part it's a good rule: Don't marry your sibling. By avoiding close family members we increase heterozygosity in our children, and everyone benefits. Close incest is a pretty extreme case, but there are other ways that our biology encourages mixing.

OF MICE AND HUTTERITES

I've already told you that smell, a much underrated sense, can give clues about the quality of a potential mate. Mice are definitely picking up on these clues. Several studies have found that mice prefer the scent of other mice that have different MHC genes. Remember, the MHC is an important part of the immune system that helps us separate self from non-self. By seeking out MHC-different mates, the mice can produce genetically diverse offspring. Their babies will have higher levels of MHC heterozygosity and probably overall heterozygosity as well. Pretty smart for mice.

It turns out that people are pretty smart too. The results of another T-shirt sniffing experiment (last one I'm afraid) confirm it. When male and female subjects were asked to rate the smells of previously worn T-shirts, they found the ones that had been worn by people with different MHCs much sexier. And the MHC-dissimilar T-shirts were more likely to remind them of former lovers. The MHC-similar shirts, on the other hand, reminded the participants of family members. Most women preferred the scent of MHC-dissimilar shirts, but not women who were on the pill. These women liked the MHC-similar smells. Which isn't surprising. Since they can't get pregnant, there's not much point in being drawn to genetically different men. Instead of sniffing out potential mates, they can just stick close to home.

This attraction to people with a different MHC makeup may explain an interesting finding among the Hutterites. But first things first: Who are the Hutterites? Related to the Amish and the Mennonites, the Hutterites are an Anabaptist group that values pacifism and communal living. Like many small religious groups, the Hutterites have a long history of fleeing persecution. They originated in the Tyrol region of Austria where their leader, John Hutter, was burned alive for refusing to renounce his beliefs. The group fled to Moravia, then Slovakia, then Transylvania, then the Ukraine, then the United States. Their globetrotting is impressive though the reasons for it are unfortunate. And the persecution didn't end once they settled in the United States. The peace-loving Hutterites refused to serve in the military, and after several clashes with the authorities, they moved to Canada in 1918 where most of them still reside. While their dress is somewhat old-fashioned (suspenders and beards for the men, dresses and polka dot bonnets for the women), the Hutterites are free to use modern technology. They even have a Web site—www.hutterites.org.

Hutterites pretty much marry other Hutterites. And, not surprisingly for an isolated religious sect, they are pretty conservative. They "strongly encourage purity in relationships," and couples usually court with the door open just to make sure. But like any other group, it boils down to boy-meets-girl, boy-likes-girl, and girl-likes-boy. It's interesting to see who likes whom. A study of 411 Hutterite couples found that the husband and wife were more likely to have different MHCs than would be predicted by random chance. Even within this closed community, it would appear that boys and girls are subconsciously seeking out genetic diversity. A separate study found that Hutterite couples who had similar MHCs were less fertile than their MHC-dissimilar peers. Maybe when your mate smells too much like you, it's hard to get in the mood.

MAKING CHOICES
THE MORNING AFTER

The Hutterites are increasing their kids' genetic diversity by making smart mate choices. But even after you've tumbled into bed with someone, it's not too late to influence the outcome. Remember that female orgasm can help "choose" the father from multiple partners. By sucking one man's sperm farther into the uterus, a woman can increase the odds that he will impregnate her. Do women "upsuck" the sperm of men who are MHC-dissimilar? I wouldn't be surprised. Other animals make similar "choices."

Female sand lizards and adders often have multiple sexual partners. That sounds pretty racy, except for the fact that they're reptiles. Scientists discovered that usually one male will end up fathering most of the offspring. It isn't necessarily the first or the last male that mates with the female. But it does turn out to be the male that is genetically most different from her. Somewhere deep inside the reptilian reproductive tract a choice is being made. One suitor's sperm is singled out as the best and is used to procreate. This male isn't the best in an absolute sense, but he is the best candidate for that particular female. Certainly his general health and attractiveness are relevant; they got his foot in the door (so to speak). But once they mate, it's his compatibility with the female that is the deciding factor. And the more distinct his DNA is from hers, the more compatible they are.

Compatibility can make a crucial difference even later in the reproductive process. Bird studies have found that the more genetically similar a bird couple is, the more likely their eggs are to not develop or not hatch. Genetic diversity is critical to producing viable offspring. Sadly, many human couples have experienced this same problem.

The desire to procreate is pretty much universal. Evolution has ensured that's the case. It's a fundamental part of our lives. When I tell you that my brother is living the American dream, you can probably picture his suburban home with a pool and white-picket fence. It's a given that said home is filled with a wife and kids. (Though I'm happy to report that there's no minivan in the driveway. Yet.) Adina and Ben, my niece and nephew, can be a real handful.

Raising Adina and Ben doesn't look easy, but happily, having them was. Well, I don't know if my sister-in-law would describe nine months of pregnancy as easy, but there were no complications. Sadly, for many couples it's not that straightforward. It can be difficult for some couples to get pregnant, while others can conceive, but can't carry a baby to term. Many spend large sums of money and experience equally large amounts of heartache in their attempt to realize the American dream.

Often technology comes to the rescue. There have been amazing advances in reproductive medicine; hormone treatments and in vitro fertilization can often deliver that long-awaited bundle of joy. But sometimes the cause of the problem is not clear. In 30 percent of the cases where couples have experienced more than one miscarriage, doctors can find no clear medical explanation. One possibility is the same kind of compatibility problem experienced by birds. Multiple studies have found that couples who have recurrent miscarriages are far more likely to be MHC-similar than control couples. The exact mechanism is unknown, but this appears to be another (albeit sad) biological mechanism that promotes diversity.

These mechanisms are often subtle, but they are there, working at different stages of the process. We can now look back at our Hutterite couples with new insight. When Hutterite man and wife are MHC-similar, they are less fertile than other couples in the same community. There are multiple possible reasons. Their MHC similarity may dampen their passion for one another. You can't have babies if you're not having sex. Or maybe they're getting it on, but Mrs. Hutterite's body isn't that excited about Mr. Hutterite's sperm since it's genetically too much like her so they're not conceiving. Or maybe they are conceiving, but the fetus isn't reaching full term. In any case, the increased sexual chemistry and fertility that genetically diverse couples enjoy are ways that our biology encourages mixing and increases genetic variation in the next generation.

But, of course, that's not the whole story.

LIKE MARRIES LIKE

There are real and powerful biological forces that encourage genetic diversity, but I've also told you that interracial marriage is still a relatively rare phenomenon. Plus, a quick look around will tell you that most couples look more similar than different. In fact, the conventional wisdom says that long-lasting couples often resemble each other. The question is, Do they start off looking the same or do they gradually come to look alike? Once again, science has the answer.

Couples usually share the same lifestyle and similar interests. If they share a passion for jogging, they'll probably both have athletic bodies. If they take up windsurfing, they'll both develop a dark tan. Living together can cause some similarities to develop. But studies of facial resemblance among couples have found that it's not just long-term partners that look alike—engaged couples do too. Since it's unlikely that they've been together long enough to morph into twins, if would appear that people seek out mates that look like themselves.

This isn't just a case of extreme narcissism. Scientists believe that we develop our internal blueprint of what a mate should look like early on in life. This blueprint is influenced by the faces that we see when we're very young. The faces that we see most often are our own and the faces of our family members. We base our blueprint on these faces and thus we're attracted to people with faces similar to our own. And so are quail.

Researchers raised Japanese quail chicks with their siblings for one month and then isolated them. At adulthood, a subject quail was put into the cleverly named Amsterdam device. Much like Amsterdam's red light district, one can stroll down a center aisle and see various members of the opposite sex offering their charms. But instead of Dutch hookers, this setup is filled with quail. Some of the quail were unrelated to the subject, some were the subject's siblings, and some were the subject's first cousins (whom they had never seen before). The subject quail spent the most time staring into the cage that held its first cousin. This was true whether the subject was male or female.

In a separate experiment, the quail were allowed to spend time together as pairs. The first cousin pairs produced eggs faster than the unrelated birds or the siblings. In these experiments, the quail recognized their siblings and were less interested in them sexually. But they were more attracted to their first cousins than to unrelated birds. Presumably the cousins fit the quails' internal mate blueprints more closely. They didn't want to mate with birds that were too similar, but too different was also a turnoff.

A creative study starring colorful snow geese shows that these mate blueprints are not innate, but are built from experiences at a young age. Snow geese are native to North America. They migrate in large flocks and come in two colors: white and blue. White snow geese prefer to mate with other white snow geese, and blue prefer to mate with blue. But when Canadian researchers played a little game of switched-at-birth, things got interesting. White snow geese raised by blue parents were attracted to blue geese later in life. And blue goslings adopted by white parents wanted to mate with white geese as adults. The biologists then took things a step further and dyed some of the parents pink. Sure enough, their offspring had the hots for pink-colored geese, even though they don't naturally occur. Finally, when goslings grew up surrounded by both white and blue geese, they matured into equal opportunity maters with no color preference.

OEDIPUS RATS

This internal blueprint has a powerful impact on mate choices. And these blueprints are not only visual. In another experiment male rats were raised by lemon-scented mothers. Yes, there are scientists out there studying lemon-scented rats. The mothers had lemon odor applied to their nipples and vaginas (How would you like to explain that job to your friends?) and then raised the baby males normally. When the male rats reached sexual maturity, they were placed with other females that were either lemon-scented or unscented, and their liaisons were videotaped. What did the researchers learn from watching this rat porn? The males mounted the lemony females more quickly and ejaculated more quickly than with the females that were regular rat scent. There are a couple of disturbing things about this study. One, people watching rats ejaculate. Two, the fact that the male rat was more sexually excited by the female that reminded him of his mother. It's a little creepy, but it makes sense. He was just following his internal blueprint.

I have a friend name Nili who only dates black men, much to her parents' consternation. They want her to settle down with a nice Jewish boy, but she can't help it; she's only attracted to black guys. All you have to do is meet Nili's dad once to realize that she isn't just being rebellious. His family comes from Yemen and, with his dark skin and curly black hair, he could easily pass for black. He's unhappy because his daughter is following a blueprint that appears to be based on him!

Sigmund Freud would have had a field day with this stuff. But I'm not saying that we want to marry our parents. I've already pointed out that close incest is a universal turnoff. I'm just saying that we build our blueprint early in life. I don't know about you, but when I was six months old I wasn't going out and drinking beer with my baby buddies. Most of the people I was exposed to were family. Thus, my family had a strong influence on the template of the future Mrs. Ziv. It's like the old song, "I Want a Girl (Just Like the Girl That Married Dear Old Dad)."

Even though our families play a big role in our early lives, they are not the only influence. Remember, snow geese that were raised in a mixed flock showed no color preferences. However, basing our internal blueprint on the people we grow up with would seem to steer us toward mates who are like us. Not mates who are different. The quail ended up with their first cousins which hardly seems ideal. Why would we evolve a system that seems to discourage mixing?

DON'T PET THE SLOW LORISES

Evolution produces animals that are adapted to their environment. The gazelle is fast because cheetahs are fast; therefore, gazelles need speed to survive in their local environment. Speed isn't innately good, and for many species it would be a waste of resources. Slow Lorises are tiny primates that have round heads and huge eyes. They're really cute, but as you can guess from their name, they're not winning any marathons. But Slow Lorises don't need to be fast; they have other defenses to protect them from predators. They live high in the treetops of the rain forest, and, surprisingly, these little fluffballs are highly toxic. They produce poison from glands in the crooks of their elbows and coat their teeth with this poison before they bite. Bottom line: no one strategy is ideal. Both fast legs and poison elbows can get the job done. You just have to be adapted to your particular environment.

This is why some animals may avoid mates that are too different. Let's say you're a quail living in the woods. Natural selection has ensured that you are well adapted to your environment. You have the right plumage to camouflage you from predators and the right foraging strategies to find those yummy seeds and leaves that you love so much. You're single and looking for some quail companionship. So you check out the prospects. There are some available quail that look like the ones you grew up with, and then there's a bird that looks totally different. If this new quail looks dramatically different than what you're used to, it may be a stranger in town and may be pretty different from you genetically. Genetic variation is good news, but is this new quail adapted to your local environment or to a different environment? Mating with the Mystery Quail will give your kids valuable hybrid vigor, but they may not inherit the coloring and instincts that they need to survive in your neck of the woods.

Polar bears and Brown bears share a common ancestor and can successfully interbreed. But even if a Brown bear decided to walk all the way to the Arctic Circle, he probably wouldn't find any interested Polar bears. The arctic is a harsh environment, and Polar bears are highly specialized in order to survive there. Extra blubber provides insulation against the cold, and the Polar bear's white coat is excellent camouflage in the arctic snow. Some claim that a Polar bear will even cover his black nose with his paw while stalking prey, to be more inconspicuous. The lack of these adaptations makes a Brown bear a bad choice for a mate. The offspring, while genetically diverse, just wouldn't be able to survive in this environment. Even if these half-Brown cubs managed to stay warm, without a white coat they couldn't catch even the stupidest of seals.

Animals have to find a balance between the costs of inbreeding and the costs of outbreeding. They want to maximize their kids' heterozygosity so they avoid mating with close relatives. But mating with someone too different can be risky too. It's critical that the offspring be adapted to the environment where they're being born. Their survival depends on it. So mates that are too different can be a turnoff. They want their babies to be able to thrive in their environment, so they base their mate on individuals that they know are adapted to their home turf—the ones they grew up with on that turf. That's fine for quail or bears, but what does it mean for people?

IT'S A DIFFERENT WORLD

When looking at evolution, and particularly human evolution, we have to consider not the world that we live in now, but the world that we evolved in. We are designed to succeed in that world, and even though our environment is very different now, our genes have yet to catch up. A classic example of this is our love of fat. We evolved in a world where food was scarce. If one did come across some fat it was good to gobble it up because fat is an excellent source of calories. That's a great strategy if you're a starving caveman, but today Twinkies and Ho Hos are just a 7-Eleven away. Now our love of fat gets us in trouble, especially in the West where obesity has reached epidemic proportions.

Similarly our system of internal mate blueprints was designed for a very different world. In the old days, we lived in relatively isolated bands as hunter-gatherers. You had contact with the people in your band, and that was usually it. You might interact with some other local tribes, but if you were an Eskimo, the only people you were hanging out with were Eskimos. Your odds of running into someone dramatically different were pretty small.

Even if you did, they might not be the best candidate for love and marriage. Let's say by some fluke, an Australian Aborigine took a wrong turn and ended up in your igloo. Sparks might fly and certainly your kids would have high levels of genetic diversity, but would they have the shorter, rounder Eskimo body that's good at conserving heat? Or would they inherit the tall, lean Aborigine physique that is designed to dissipate heat? There's not much point in having beautiful, symmetrical kids if they freeze to death.

Of course, the reverse is true too. An Eskimo who wanders into the Outback probably wouldn't be an ideal parent candidate. There's too great a risk that half-Eskimo offspring will grow up without the Aborigine body and dark skin that evolved over hundreds of generations in the hot Australian sun. These traits evolved for a reason. Babies are cranky enough without heat stroke and skin cancer to worry about.

There was no point in being attracted to dramatically different people in the old days because you would never meet any. Even if you did, there was too great a risk that their genes wouldn't be a good fit with your local environment. It made a lot more sense to be drawn to people you knew would have the right adaptations. These internal blueprints, based on the local community, were a good way of doing that.

But we don't live in that world anymore. Mass movements of populations mean that many of us are no longer living in the environment for which our genes were designed. It doesn't matter if our genes don't fit out environment because we've been able to use technology to make up for it. With GORE-TEX snowsuits and central heating, anyone can live in Alaska. Similarly, the Outback isn't just for Aborigines anymore, thanks to air conditioning and sunscreen. Yet we still pick mates based on the people we grew up with. People who are often genetically similar to us.

Just like our fat-loving appetites, this strategy is no longer ideal. For the modern world, it would be far better to seek out the most genetically distinct mate possible. And it would be far better if I liked celery more than Ding Dongs. Sadly, it's not up to us. Evolution can be a slow process, and we are still saddled with the genes and strategies of our ancestors. However, in the next chapter I'll tell you how our changing world is about to turn Like Marries Like completely on its head.

IN THE MEME OF THE FATHER

So, some biological forces promote mixing while others discourage it, but that's not the whole story. When it comes to picking a mate, society has just as much to say as biology. Society and biology are often depicted as polar opposites, but in *The Selfish Gene*, Richard Dawkins points out that social ideas can spread and evolve just like genes. He names these units of cultural evolution, "memes." He probably didn't realize it at the time, but the concept of memes has itself become a powerful meme. It has spread widely and spawned an entire field: memetics. There is no clear definition of a meme, but any idea that can spread from one person to another may be considered a meme. A joke, an advertising jingle, or a political movement can all be memes. And just like genes, some memes are more successful than others. Some of the oldest and most success-ful memes are religions.

My father is a very logical man. An engineer by trade, he's always saying things like, "If it's worth doing, it's worth doing right." So if a man approached him and said, "Good news! I just talked to God and we can be his chosen people. But we all have to give up bacon forever and cut off a piece of our penises," I'm pretty sure my dad would call the cops. And yet, when I was eight days old he made sure I was circumcised, and when we have breakfast, the only thing on the plate with his eggs is hash browns. Such is the power of religious memes.

It's not by accident that religious memes have had such staying power and such a strong influence on the lives of their followers. Most religions have common elements that make them both appeal-ing and commanding. In general, religion offers a set of rules to live by. Those who follow the rules are rewarded, and those who don't are punished. The rewards are extravagant: different versions of everlasting life in heaven, sometimes with beautiful virgins to keep you company. The punishments are equally horrific: burn-ing in hellfire for all eternity, for example. I think one of the most appealing aspects of religion is this sense of justice; good deeds are recognized and the bad guys get what's coming to them. This is especially attractive in world where justice is often missing from regular life.

In discussing religious memes, I'm going to focus on Judeo-Christian faiths. Not because those religions are somehow special, but because they are very successful memes that most people are familiar with. Many of these religions' rules ensure the survival of the religion itself. The importance of the religion is paramount. After all, the first commandment is "I am the Lord thy God. Thou shalt have no other gods before me." A religion that discourages competition is going to fare better than one that allows it. This is why faith is such an important component of religion. In an ingenious twist of logic, one of the rules is that you aren't allowed to question the rules. Just ask Thomas.

When some of the other apostles tell Thomas that they saw Jesus rise from the dead, he is skeptical, "Except I shall see in his hands the print of the nails, and put my finger into the place of the nails, and put my hand into his side, I will not believe." He wants to see for himself. He gets the opportunity eight days later. Thomas sees Jesus and finally believes that he has been resurrected. But Jesus scolds him for requiring proof, "Because thou hast seen me, Thomas, thou hast believed; *blessed are they that have not seen, and have believed.*" I added the emphasis to highlight the moral of the story: Thomas was wrong to be skeptical; blind faith is good. Christian schoolchildren for thousands of years have been warned not to be a Doubting Thomas.

In addition to having faith, one of the most important things a follower can do is instill that same faith in his or her children. This guarantees that the meme will continue in the next generation. Many religions also encourage their practitioners to have as many children as possible, which ensures more followers every generation. Of course, in order to make sure the next generation is instructed properly, both parents need to be followers. I've already mentioned that in Deuteronomy those who intermarry are threatened with God's wrath. This is not the only time mixing is prohibited. In the

Book of Malachi we learn that, "Judah hath dealt treacherously, and an abomination is committed in Israel and in Jerusalem; for Judah hath profaned the holiness of the Lord which he loved, and hath been intimate with the daughter of a strange god." And the punishment for this "abomination" is "The Lord will cut off the men that doeth this." Some scholars interpret "cut off" to mean shunned, but many believe it means that the offenders will be killed.

Jezebel is often depicted as the most wicked woman in the Bible, but her only crime was worshipping a different God. When King Ahab married her, he "did more to provoke the Lord God of Israel to anger than all the kings of Israel that were before him." Elijah later prophesizes, "The dogs shall eat Jezebel by the wall of Jezreel"—a prediction that comes true after she is killed by being thrown out of a window. That's pretty harsh.

There are many examples of this. During my research I found a Ku Klux Klan Web site that claimed to list over two hundred Bible verses that prohibit racial mixing. Of course I'm not trying to put religion and the Klan in the same category. It's not my goal to attack religion, merely to point out that religious doctrine has been a significant barrier to interracial marriage over the last few thousand years. This wasn't the goal of religion, more of a side effect. Religions that were strict about intermarriage didn't get diluted by other faiths—they were more successful. Permissive religions are like slow cheetahs, they don't last.

Certainly religion is not the only cultural factor that can influence marriage decisions. Racism is a powerful meme that can reduce the likelihood of intermarriage. Bigots are unlikely to hang out with people they consider inferior, much less date them. And racism extends further than just the racists themselves. You may be perfectly liberal about race, but if your dad has a white hood in his closet you're going to be very careful about whom you bring home. There's also the opinion of the community at large. Countless parents have asked their kids, "But what will the neighbors think?"

Your culture can also encourage marrying someone like you

without being explicitly racist. In Chapter One, I discussed the story of Tevye, who wants his daughters to marry someone like them. Part of his desire comes from religion—he wants his daughters to marry Jewish men. This will ensure Jewish grandchildren. Sometimes people feel more comfortable with spouses and in-laws that are similar to them. Not just in religious background, but level of education, socioeconomic status, language, and so on. The easiest way to accomplish this is to stay close to home when choosing a mate. Some people stay really close to home.

KEEPING IT IN THE FAMILY

I mentioned before that in parts of the Middle East and Africa certain types of inbreeding are preferred. Not close incest, of course. There is no society in the world that encourages parent-child or brother-sister marriage. But cousin marriages are viewed very differently in different cultures. In the West, first cousin marriage is usually seen as backward and unnatural, but in the Middle East it is not only accepted, but also it's seen as the ideal. In traditional Arab society, the model marriage is between a man and his father's brother's daughter. Basically, his paternal first cousin. This has a couple of advantages: 1) the woman doesn't have to get used to a new last name, and 2) it results in a very strong family unit.

It's interesting that not all cousins are created equal. The preferred wife or *bint'amm* is the paternal cousin of the husband. In the Middle East, where the men hold the political power, you want to be able to count on the support of your brothers, sons, and so forth. This type of cousin marriage consolidates that support instead of fragmenting it. There's no danger of being torn between your uncle and your father-in-law because they're the same man. Thus brothers, sons, uncles, and nephews are united by multiple familial ties. These ties bind more strongly than any other. In fact, some political analysts believe that understanding this type of family unit is critical to understanding the politics of the Middle East.

It's possible that nation-building attempts by the West have failed there because family and clan ties are so much more important than national ones.

Though it's more formalized in the Middle East, marrying someone like you reduces conflict almost anywhere. Sometimes it's just easier to be with someone who grew up with similar experiences to yours. This doesn't just apply to romantic relationships. Often people prefer to be friends with others who speak the same language, eat the same food, and celebrate the same holidays. For this reason, even when they move out of their native land, people often settle in ethnic enclaves. This can also be a barrier to mixing. Even in a brand-new country, halfway around the world, you and the boy or girl next door may still be from the same village. This tendency of immigrants to build ethnic communities has been dubbed the "salad bowl" effect. Instead of a melting pot where different groups blend together into one homogenous soup, the salad bowl model illustrates how different populations can live together, but retain their distinct identities. A crouton can join the salad and still remain separate, maintaining its individual flavor and texture. Only time will tell if America's future is that of soup or salad.

WHAT'S NEXT?

I like the idea of cultural evolution, but the parallels only go so far. One important way that genes and memes are different is that memes can move much more quickly. Genes are passed from one generation to the next, and it takes time for their frequencies to change. Memes, on the other hand, can spread from person to person and can appear and disappear in the blink of an eye. (Remember tamagotchis?) Societal ideas about interracial marriage are unlikely to change that quickly, but they have come a long way in the last couple of generations.

In 1944, Gunnar Myrdal, who later won the Nobel Prize in Economics, published a study on race relations in the United States called *An American Dilemma: The Negro Problem and Modern Democracy.* In it, he wrote that "even a liberal-minded Northerner of cosmopolitan culture will, in nine cases out of ten, express a definite feeling against" interracial marriage. Nine cases out of ten may have been an optimistic estimate. A survey in 1958 found that only 4 percent of white Americans approved of interracial marriages. By 1994, that number was 45 percent—a more than tenfold increase.

Even if historically cultural divisions kept us apart, they seem to be abating. Multiculturalism and multiracialism not only are more accepted, but also have become downright trendy in some quarters. What does the future hold for race mixing in our society? Find out in the next and final chapter.

7

Is the Pot Melting?

THE FUTURE OF RACE IN AMERICA

IN THE MOVIE *Bulworth*, Warren Beatty plays a suicidally depressed senator named Jay Bulworth. He's about to face financial ruin so he decides to end it all and puts out a contract on his own life. Surprisingly, this movie is a comedy. His impending death frees Bulworth; he no longer has to worry about polls, fund-raising, or reelection campaigns. For the first time he's free to speak the absolute truth, and the truths that come out of his mouth are both shocking and hilarious. It's an edgy and daring political satire in which Bulworth offends blacks, Jews, and the media with his brutal honesty. He also raps, begins an affair with Halle Berry, drives his handlers crazy, and does it all while swearing up a storm. Supposedly the f-word is uttered ninety-five times.

It's entertaining to imagine what a politician would say if he spoke the unfiltered truth and Warren Beatty rapping is a sight to see, but *Bulworth* is more than just a comedy. It's also full of scathing social commentary about politics, the media, and race relations in America. When discussing race in an interview, Bulworth says, "All we need is a voluntary, free-spirited, open-ended program of pro-creative racial deconstruction." He goes on to explain: "Everybody's just got to keep fucking everybody 'til they're all the same color." I warned you about the f-word.

"Procreative racial deconstruction" is a funny concept, but it didn't originate with Bulworth. In 1963, then liberal writer Norman Podhoretz published an essay titled "My Negro Problem—And Ours." Podhoretz took a hard look at black-white race relations in America and decided they were unfixable. Podhoretz wrote that these problems could not be solved while race itself still existed: "That means not integration, it means assimilation, it means—let the brutal word come out—miscegenation." This was a very bold statement. Especially in 1963—four years before the Loving decision guaranteed the right of interracial couples to marry—miscegenation really was a bad word.

These sentiments don't just come from the left. More recently, Douglas J. Besharov, a scholar for a conservative think tank called the American Enterprise Institute, has said that interracial children may be "the best hope for the future of American race relations."

I Have a Dream

In 1963, the same year that Norman Podhoretz went public with his "Negro Problem," Martin Luther King Jr. stood at the Lincoln Memorial in Washington, D.C., and delivered one of the most famous speeches in American history. In it he said:

> And so even though we face the difficulties of today and tomorrow, I still have a dream. It is a dream deeply rooted in the American dream.
>
> I have a dream that one day this nation will rise up and live out the true meaning of its creed: "We hold these truths to be self-evident, that all men are created equal."
>
> I have a dream that one day on the red hills of Georgia, the sons of former slaves and the sons of former slave owners will be able to sit down together at the table of brotherhood.

> I have a dream that one day even the state of Missis-
> sippi, a state sweltering with the heat of injustice, swel-
> tering with the heat of oppression, will be transformed
> into an oasis of freedom and justice.
>
> I have a dream that my four little children will one day
> live in a nation where they will not be judged by the color
> of their skin but by the content of their character.
>
> I have a dream today!

Just reading his words gives me chills. Martin Luther King Jr. was a great speaker, but his powerful words make it clear that he was also a great optimist. He believed that we, as a nation, could get past hundreds of years of racial discord and live in harmony. Podhoretz and Besharov and Bulworth have given up on that dream. In a way, they sound like stern parents: "You Americans can't play nice with race so the only solution is to take it away!" This attitude would be funny if it weren't so pessimistic. Much like those scientists who are afraid to acknowledge ethnic differences, they seem to think that racism is an inevitable consequence of race. Mixing race to the point of homogeneity is the only way out.

I disagree. No matter how bleak the state of American race relations (and at times it looks very bleak indeed) I'd rather align myself with a message of hope like Dr. King's than a message of resignation and despair. I also think that the idea of eliminating race entirely is a little naïve. Prejudice is a forgone conclusion. Some people want to feel superior to others; the criteria that they use to accomplish that, be it skin color, level of education, income, or upbringing, are secondary.

Brazil has a long history of interracial coupling, and the diversity of its citizens reflects that. Brazil hasn't eliminated race or racism; they've both just become more of a continuum. In 1976, the Brazilian Institute of Geography and Statistics conducted a census that listed an incredible 134 choices for skin color! The options included "Bem-branca" (very white), "Canela" (cinnamon), "Café-com-leite" (coffee with milk), and "Burro-quanto-foge" ("donkey running away," meaning unknown racial origin). The more branca (white) the better. An old Brazilian saying breaks down the desirability of various women: "Branca for marriage, mulata (mixed) for sex, preta (black) for work." So much for intermarriage eliminating racism.

I disagree with the Bulworth theory that race mixing is the key to race relations. But I do agree that mixed marriage is the future of this country. Not because it will fix any of our problems as a nation and not because it's the right thing to do. And not because I'd like to live in a color-blind society. Interracial marriage is the future, not because it *should* be, but simply because it is. It's inevitable. Why? I've already told you why. Because of snow geese.

In the last chapter I told you that, early in life, snow geese build a blueprint of what their future mate should look like, just like people do. Snow geese raised by blue parents wanted to mate with blue geese. Those raised by white parents looked for a white mate later in life. But goslings raised in a mixed flock had no color preference. I'm going to say that again: *goslings raised in a mixed flock had no color preference.*

This is huge. This means that as our flocks get mixed up and our communities become more diverse, our blueprints will become more general. Children that are exposed to a variety of ethnicities at a young age will most likely be attracted to a variety of potential mates when they grow up. A diverse upbringing means that like will no longer exclusively marry like. But do humans work the same way as geese? For the answer, let's take a look at people who have been raised in a mixed flock.

WELCOME TO THE HOTEL CALIFORNIA

California is one of the few states that have no ethnic majority. It's also home to more interracial couples than any other state in the union. After the 2000 Census, I looked at a list of the metropolitan areas with the highest percentage of interracial Americans. Of the top thirty most interracial cities, twenty-one are located in California. Sounds like a mixed flock to me. How has this affected its rate of intermarriage?

Living in a multicultural and multiracial mecca like California should have a big impact on the number of interracial marriages. Californians regularly come in contact with people of every color and creed. One, this means they have the opportunity to date across racial lines. And two, living in a mixed flock should broaden the average Californian's mate blueprint. With no color preference, Joe California should be more of an equal opportunity dater, which means more mixed marriages. Let's take a look at the numbers.

The Public Policy Institute of California analyzed birth records from 1982 to 1997 in order to get a picture of trends in the multiracial population. In 1982, about 12 percent of all children born were multiracial. Fifteen years later that number had jumped to about 14 percent. That's a significant increase (about 15 percent), but when I first saw these numbers I was disappointed. Fifteen years is more than half a generation. I was sure that the mixed flock phenomenon would reduce preferences for marrying within one's group. I was confident that over fifteen years this change in preferences would have a huge impact. Where was the big jump in mixed relationships, and thus mixed babies, that I predicted? Could I be wrong? Before I admitted defeat, I took a closer look at the data.

One great thing about this study is that they also recorded the birthplace of the mother. Obviously, all these babies were born in California, but some of the mothers were native to the Golden State while others had immigrated to the United States and settled in California. This is a significant distinction in a place like California. Everyone wants to live in the land of hot sun and hot women. And as a result, California has a large annual influx of immigrants. In fact, by 1997 almost half of the births in California were to foreign-born mothers. When we look at the native-born versus foreign-born mothers, a fascinating trend emerges. In 1982, mothers native to California gave birth to multiracial babies 14 percent of the time. By 1997, that had increased to about 21 percent, a 50 percent increase.

More than one-fifth of the babies born to native Californians were interracial. There's the big jump I was looking for!

But what about foreign-born mothers? At the beginning of the study, they had multiracial babies less than 8 percent of the time. Fifteen years later, in 1997, that number had changed to 7 percent of the time. It had hardly changed at all; in fact, it dropped slightly. The big jump in multiracial births did occur, but only for native-born mothers. It was masked in the overall statistics by the foreign-born mothers. These immigrant mothers were responsible for an increasing number of births over the course of the study, but they were no more likely to intermix as time passed.

No doubt many of the foreign-born mothers were already married when they immigrated, which helps to explain why they are unlikely to give birth to interracial babies. In some cases a language barrier may also have prevented mixing. But I think that we also need to look at the very different flocks that the two types of mother grew up in. A foreign-born mother most likely grew up in a more homogeneous environment. She developed a very specific blueprint for her future husband. That blueprint outlined a man of her ethnicity. Even given the opportunity to intermarry in a diverse place like California, foreign-born women are uninterested. They are not attracted to men of other ethnicities because they don't match the blueprint.

The native Californian mothers, on the other hand, grew up in a mixed-flock environment. Their internal husband blueprints were influenced by the hodgepodge of races and cultures present in California. When the time comes to find a husband, these women, with their more general blueprints, are far more likely to be attracted to and marry a variety of men than immigrant mothers. Ironically, immigration, which is absolutely necessary for intermarriage to occur, can also be a barrier to it. But it's just a matter of time. The immigrants may not mix, but their children will. Estimates for Hispanic immigrants show that only about 8 percent intermarry. For the second generation, however, that number jumps to 32 percent. By the third generation it's 57 percent. The immigrants may not be happy about it, but their grandchildren are more likely to intermarry than not.

WHAT ABOUT THE REST
OF THE COUNTRY?

In 2000, for the first time, the U.S. Census allowed respondents to check more than one box for race. This is an important acknowledgment that the interracial community is growing, and many Americans don't fit neatly into a single category. It also gives us a better picture of the interracial population in the United States. Of course, this picture isn't perfect.

I've repeatedly argued that race is a continuum and so I've avoided specifically defining racial categories, but the U.S. Census Bureau doesn't have that luxury. In order to capture the racial makeup of the country, they have to decide what counts as a race and what doesn't. For the 2000 Census, respondents had six general choices for race (again they could check more than one): American Indian or Alaska Native; Asian; black or African American; Native Hawaiian or Other Pacific Islander; white; and other. Of course, even while attempting to collect racial data on more than 280 million Americans, these bureaucrats were careful to follow the party line and say that they don't believe in race either: "The concept of race as used by the Census Bureau reflects self-identification by people according to the race or races with which they most closely identify. These categories are sociopolitical constructs and should not be interpreted as being scientific or anthropological in nature."

The U.S. Census also wanted to capture the Hispanic population demographics of the country, but, according to the Bureau, "Hispanics and Latinos may be of any race." This seems reasonable. Sammy Sosa is obviously "Black or African American," but Spanish is his native language and he was born in Latin America, in San Pedro de Macorís, Dominican Republic. To the people at LatinoSportsLegends.com, and many others, he is clearly also

Hispanic. To deal with this issue (not Slammin' Sammy specifically, but people like him), the Census Bureau added a separate question concerning the respondent's "origin": "Origin can be viewed as the heritage, nationality group, lineage, or country of birth of the person or the person's parents or ancestors before their arrival in the United States." That definition is a little confusing, but the basic categories for origin make things much more clear. The choices are Spanish/Hispanic/Latino and Not Spanish/Hispanic/Latino.

Unfortunately this seems to have led to some confusion. About 20 percent of Americans who checked off more than one racial category were of Spanish/Hispanic/Latino origin and selected white and other for race. That, in itself, is not a problem, but the other line has a blank space for explanation. After selecting other many of these respondents wrote in "Hispanic" or "Latino." This seems to indicate some confusion about the difference between the racial categories and the question of origin. This may mean that some of the Americans counted as interracial are not actually of mixed race. That is a problem. And it's not the only one.

There's also the case of my friend Karen. Karen's father is Russian and looks it. He has very fair skin and green eyes. Her mother, on the other hand, is Yemeni and looks about as different from him as chalk and cheese. Karen's mom has black hair, brown eyes, and skin that on the 134-point Brazilian scale would probably be somewhere between Canela (cinnamon) and Chocolate (chocolate brown). As a small child, Karen would sometimes become confused because her maternal grandfather looked so much like Bill Cosby.

But when the 2000 Census arrived in her mailbox, Karen unhesitatingly checked off "white" and mailed it back in. She's not trying to hide her mixed heritage. On the contrary she's quite proud of it, but she's never felt that this type of survey gave her the opportunity to express it. There is no clear category for her Middle Eastern ancestry, and for years most forms of this type only allowed one answer and explicitly stated "white (includes Middle East)." Karen got used to just checking off the white box and that's what she did in 2000. There's no doubt in my mind or hers that she's interracial, but she was recorded simply as white.

Bottom line: there are problems with the U.S. Census. In some cases the mixed group numbers may be exaggerated; in others they may be underreported. Of course, since race is really a continuum, there is no ideal way to define or document it. The data are not perfect, but with those caveats in place, let's take a look at the 2000 Census.

THE GREAT DIVIDE

When I first started digging into the 2000 Census data, I wasn't expecting the numbers to be the same everywhere. I knew that there were bound to be more interracial people in some parts of the country than in others. Specifically I expected to see a divide between the North and the South. Maybe that was unfair of me. The Civil War has been over for 140 years, after all. But I still thought that interracial births would lag behind in the Southern states, especially since interracial marriage was illegal in most of them until 1967. But I was wrong. Not about the South—interracial people are still a rarity there—but about the North. The large interracial gap that I was expecting does exist, but it isn't between the North and the South. It's between the East and the West.

Despite being the home of abolitionists and having a reputation for liberalism on race, the Northeast's numbers of interracial citizens are only slightly higher than the South's. It's the western United States where most of the interracial baby boom seems to have taken place. This is clear from the numbers, but even clearer from a color-coded map of America. It looks almost like two different countries, with the Mississippi River as the border. Of the top ten most interracial states, all are in the West except New York, which came in at number eight. Of the bottom ten, all are in the East.

The farthest west of the bottom ten is Iowa, at number forty-two, which is just barely on the west side of the Mississippi. Of course, the percentage of interracial Americans isn't that high in the West; it just looks that way when compared with the East. The tenth most interracial state is Arizona; 2.86 percent of Arizonans checked more than one box. That's really low! But it's still over two and one-half times as many as Tennessee, which ranks fortieth with 1.11 percent. My initial instincts weren't completely off; many of the Southern states are at the bottom of the list, but they're not alone. They're keeping company with the Northeast. Maine is ranked forty-eighth and New Hampshire is forty-fourth.

What is it about the West? Why is the west side of the pot melting so much more than the east? Part of the answer may come from one of the problems with the Census that I told you about. The West has a far larger Hispanic population than the East. If confusion resulted in Hispanics overreporting their interracial numbers, the West's statistics would be inflated more than the East's. No doubt that's a factor, but I don't think it's the whole story. After all, 12.31 percent of Illinois's population is Hispanic, ranking it tenth in the country, but its interracial numbers do not stand out. It's firmly in the middle of that pack at number twenty-five.

What else could it be? The frontier spirit? The warm weather? I think what sets the West apart is a simple consequence of its geography—it's in the west. That sounds redundant, but it's sig-nificant because Europeans started colonizing this continent from the east. What I'm saying is that it's the newness of the West that makes the difference. Most of the cities of the West barely existed a hundred years ago; they are the products of recent, rapid growth. Growth that continues to this day. The top five fastest growing states from 1990 to 2000 were all western: Nevada, Arizona, Colorado, Utah, and Idaho. I think the newness and rapid growth of the West has resulted in big cities that are less burdened by historical racist baggage. Cities that are more integrated. Cities that are more like mixed flocks.

WHO'S MIXING AND WHO ISN'T

This same Census data can give us some insight into exactly how mixed these flocks are. By analyzing how segregated two groups are in a metropolitan area, we can calculate what's called a Dissimilarity Index. This number reflects how integrated the two groups are. A Dissimilarity Index of 0 means that the two groups are completely integrated. A score of 100 means they are completely segregated. We'll start with whites and blacks.

I know that whites and blacks often live in separate communities, but I was still stunned when I looked at the fifty American cities that are most segregated with respect to whites and blacks. The ignoble distinction of being number one goes to Gary, Indiana, with a mind-boggling Dissimilarity Index of 87.9! The approximately 430,000 white and 120,000 black citizens of Gary appear to be living almost completely separately. I wonder if the white people in Gary are even aware there are black people in their city. A depressing beginning, but going down the list an interesting trend emerges: it's overwhelmingly eastern. The western most city on the list is Houston, Texas, at number forty-five. The list of the fifty most segregated cities with respect to whites and Asians shows a similar east-heavy pattern.

However, the white/Asian numbers are significantly better. The city where the white and Asian populations are least integrated in the country is Ann Arbor, Michigan, with a Dissimilarity Index of 64.0. That's not great, but it's a hell of a lot better than the white/black situation in Gary, Indiana. In fact, there are 113 cities in the United States where the white/black Dissimilarity Index is higher than 64. That means 113 communities where the white and black populations are more segregated than the most segregated white/Asian city in the nation! Bottom line: whites and Asians are mixing more than whites and blacks.

My mixed flock theory predicts that Asians should be more likely to intermarry because of this. And they are. In 2000, 14 percent of respondents who checked Asian also checked another box. Fourteen percent of Asians in this country are interracial, compared with a mere 5 percent of blacks. Segregation is the obvious culprit, but the reasons behind it are also significant. Segregation could just be a historical holdover, or it could be a sign of lingering racism. If it's just a historical holdover, then living in separate flocks has led to more specific mate blueprints which has led to people marrying their own kind. If cities have remained segregated because racism still exists, then we have the problem of specific blueprints and the additional barrier of racial prejudice. Either way, segregation is a major obstacle to intermarriage.

The fact that whites and Asians are less segregated also means they have more opportunities to meet and fall in love. And for those who grew up in this country, being surrounded by a mixed flock means a more open mate blueprint. The numbers here are particularly impressive. In 1990, an incredible two-thirds of Asians in their twenties born in the United States married outside their race, mostly to whites. Growing up in American cities, interacting with white people on a daily basis, has clearly affected their internal mate blueprint. Gone is the Asian-specific blueprint of their parents, replaced with a more general one. At the very least, a white or Asian mate can fit the spouse mold—the number of Asians marrying people other than whites or Asians is still quite low.

Not all Asians are equally likely to intermarry. I told you in Chapter One that black men are far more likely to marry whites than black women. The reverse is true for Asians in this country. Asian women marry white men far more frequently than Asian men marry white women. About 75 percent of all white/Asian marriages involve an Asian wife and a white husband. There is a similar trend for unmarried couples that live together. Some believe that this imbalance is caused by the fetishization of Asian women as subservient and seductive geishas. Others have argued that white men are perceived as more masculine which makes them more desirable. Regardless of the reason, I would expect to see some bitterness from Asian males who appear to be left out in the cold.

The gender gap for black/white and Asian/white marriages shows that many factors determine intermarriage rates. In general, however, integration is critical. We see more intermarriage wherever we see more integration: in the West and in large urban areas like Honolulu, Hawaii (14.9 percent interracial); the Bronx, New York (5.8 percent interracial); and Portland, Oregon (4.2 percent interracial). We don't see intermarriage in rural areas in the East that are far more homogenous such as Madison County, Mississippi (0.5 percent interracial) and Luzerne County, Pennsylvania (0.6 percent interracial).

GET READY FOR A MIXED RACE EXPLOSION

I believe we are on the cusp of a mixed race explosion. All of the pieces are in place, and all of the factors that encourage intermarriage are on the rise. The white majority is less of a majority than ever before. In 1980, 79.57 percent of Americans were white. In 2000, the number had dropped to 69.13 percent. This was accompanied by significant growth in the populations of other races—notably Asians and Hispanics. The Asian population more than doubled in this twenty-year period; the Hispanic population almost doubled. A smaller white majority means greater diversity and more opportunities for intermarriage.

The overall population of the United States has also grown and in all the right ways to boost interracial marriage. The West, which we've established as the epicenter of America's multiracial boom, is growing faster than any other region. Its population increased by 19.7 percent in the 1990s, and more Americans are living in urban areas than ever before. City-dwellers are far more likely to intermarry, and in 2000, four out of every five Americans lived in a city; almost one-third of Americans lived in a big city with a population of at least 5 million. Compare that with a hundred years ago, when more than 60 percent of America lived in rural areas.

With all of these factors coming together, mixed marriages will skyrocket. As more immigrants arrive on our shores, as more Americans live in big cities, especially cities in the West, as our communities become more and more heterogeneous, society will become increasingly accepting, and the pot will melt with increasing ferocity. Our children will grow up in a multicultural and multiracial smorgasbord, not just on the streets outside their door, but on television, in movies, and in magazines. The decreasing white majority means more and more minority consumers hungry for content. Their clamoring and their purchasing power will prove impossible to ignore. This surge of interaction with other races, both in the real world and through the media, will have a direct impact on the mate blueprints of the next generation. And once they grow up and start seeking out mates, their more cosmopolitan blueprints will ensure that they have more and varied options. This, in turn, means future generations of Americans with unprecedented racial and genetic diversity. Bulworth would be so pleased.

It's already begun. Only 2.4 percent of Americans were interracial according to the 2000 Census. I've already told you that these numbers were not even distributed geographically. There are significant trends not just across space, but also across time. Only 1.9 percent of adults were interracial, but of Americans under the age of eighteen, 4 percent checked more than one box. That means that children, the next generation of Americans, are more than twice as likely to be mixed as the current generation. This is especially true for groups like Asians that have immigrated recently. In Maryland, 8 percent of adult Asians are interracial, compared with 20 percent of Asian children. These are significant numbers. Twenty percent of Asian children in Maryland are mixed; 21 percent of babies born to native Californians are multiracial. These numbers cannot be ignored. An interracial explosion is coming. It won't happen overnight, and it won't happen everywhere at the same time or at the same rate. But it's coming. And it's coming soon. And it will change the face of America.

AND WHAT OF THE TEVYES?

In Chapter One, I claim that the Tevyes, the parents who want their children to marry someone like them, are the most significant barrier to interracial marriage. Not all members of the current generation are Tevyes; many would be happy with whomever their child brought home. But I think there are many Tevyes still out there, and they will fight against intermarriage as hard as they can. They will fight, but in the end they will lose. Just as the original Tevye did. In *Fiddler on the Roof*, Tevye is horrified that his daughter wants to marry a non-Jew. He forbids the marriage and threatens her with the most powerful weapon he has: disownment. But he can't stop her. She marries the man she loves, and Tevye declares that she is dead to him.

It's a tragic ending, but I admire Tevye's daughter. Not because rebellion is commendable, but because she had the courage to do what she thought was right. I think she made the right decision. Not because she married a man who was genetically different, but because she married a man that she loved. As I've said, this book is not an instruction manual. I'm not telling you to pick out your future spouse based on racial criteria. To me, the idea of seeking out a mate who is racially different from you is as silly as seeking out one that's racially the same. If I were to give you instructions, they would probably be something like "Follow your heart."

Anyway, I don't need to instruct people to seek out interracial unions because they are inevitable. America will fulfill its destiny as The Melting Pot regardless of what I or anyone else has to say about it. I know that the children of these unions will have significant advantages when it comes to looks, health, athleticism, and fertility, but that's just gravy. I'm excited about these advantages, but I doubt that anyone will pick out a spouse with them in mind. These interracial advantages, human hybrid vigor, are not a goal, but just a natural consequence of living in a smaller, more integrated world.

GHOST IN THE GENE MACHINE

I'm in awe of the power of genetic diversity. I think it's truly mind-boggling that on average, interracial people can grow faster, run faster, and heal faster. Not to mention bring their partner to orgasm faster. I know that the genetic advantages of mixed race individuals are real and powerful. I've tried to convey that awe to you. I also know that focusing on biology can feel cold or even Machiavellian. Are we just gene machines following our programming, trying to pass our DNA on to the next generation? It's kind of depressing to think we are just the sum of our genes and nothing more. Where is the "triumph of the human spirit" that we're always hearing about in movie trailers?

Speaking of movies, the science fiction thriller *Gattaca* delves into exactly this topic. It asks the question, "What makes us who we are?" The film portrays a futuristic world where children are genetically engineered to contain only the best of their parents' genes. All negative genes are eliminated in order to produce "superior" children. The main character, Vincent, played by Ethan Hawke, is conceived the old-fashioned way (in the backseat of a car), with no genetic intervention. Like any of us, he is a random combination of his parents' genes, good and bad. He suffers from a heart condition among other problems.

His parents learn their lesson, and when his younger brother, Anton, is born, it's through the new, "better" methods. Vincent works hard, but struggles to succeed in a world where he is constantly judged on his DNA. Anton inherits the best possible combination of their parents' genes; by all rights he should be genetically and physically superior. Their parents could have had a thousand kids the normal way, and none would have been as perfect as Anton. In the climax of the film, Vincent challenges Anton to a contest of physical endurance. In a dramatic midnight confrontation, they go to the beach and compete to see who can swim out the farthest. The smart money is on Anton and his carefully selected DNA, but in the end it's Vincent who triumphs. Anton can swim no farther, and Vincent must carry him back to shore. Later Anton is shocked that he lost to his genetically inferior brother. Vincent says, "You want to know how I did it? This is how I did it, Anton: I never saved anything for the swim back."

I love that line! It's a classic example of "triumph of the human spirit." Stuff like that is the reason we go to the movies. But how often does the human spirit triumph in real life? *Gattaca* is a new spin on an old debate: nature versus nurture. Like most black-and-white issues, I think the answer here is gray. Even though I'm a biologist, I enjoyed the dramatic ending to *Gattaca*, because I know that genes aren't everything.

I've always been fascinated by professional athletes. Of the millions of kids who dream of growing up and playing in the big leagues, only a handful will make it. They are truly the best of the best. Certainly these players need the strength and reflexes that can only come from their genes, but none of them made it by DNA alone. If there's any common thread among these athletes, it's discipline, dedication, and near-obsessive training. Just ask Ted Williams.

It's often argued that Ted Williams was the greatest hitter in baseball history. And no wonder, he was the American League MVP twice, led the league in batting six times, and was a two-time Triple Crown winner. Most impressively, Williams was the last player to hit more than .400 in a season. That means that he was able to do what he described as "the hardest thing to do in sports"—hit a baseball—and he was able to do it consistently, 40 percent of the time. No other player has accomplished such a feat since Ted did it in 1941.

Williams wasn't just coasting on raw talent; he is often described as an obsessive student of batting. One of Ted's roommates was rudely awakened one night when something hit his bedpost with a loud crash, causing his bed to collapse. Ted was doing practice swings at three A.M., in the middle of the room, in the dark. He was even known to practice swinging while he was in the outfield and the other team was at bat. He described himself as "a guy who practiced until the blisters bled, and then practiced some more." Ted's book, *The Science of Hitting,* is still considered definitive.

It would appear that Ted applied this same work ethic to everything in his life. He loved fishing, and later in life he became one of the best sport fishermen in the world. He is a member of the Fishing Hall of Fame! Earlier in his career, Ted missed several seasons of baseball because of military service during World War II and the Korean Conflict. Ted served as a Marine pilot and flew thirty-nine missions in Korea as John Glenn's wingman. When asked about Ted Williams's flying ability, the future astronaut described him as, "The best I ever saw."

Of course, genes aren't everything, but they are something. A very important something. I don't think Ted Williams would have been baseball's greatest hitter without all of his hard work. But he didn't make it on hard work alone. Without the right combination of genes, Ted wouldn't have had the raw talent that allowed him to mold himself into a hitting machine. One of Ted's genetic gifts was his eyesight. His superior vision was legendary, and many thought it was a big component of his hitting prowess. Ty Cobb took a break from beating up fans to say, "Ted Williams sees more of the ball than any man alive." Williams liked to downplay his visual abilities, but when he entered the Navy, his eyesight was measured as 20/10. It wasn't practice that allowed Williams to see twice as far as the average person. It was great genes.

The vast majority of people don't have the kind of raw talent necessary to play professional sports. Sadly, I'm one of them. I like basketball, but no amount of practice will make me good enough to play in the NBA. I just don't have the height. Or shooting ability. Or rebounding ability. Or defense. Bottom line: I just don't have the genes for it.

I wasn't surprised to learn that Tiki Barber, one of the best players in the NFL, has an identical twin brother, Ronde, who is one of the best cornerbacks in NFL history. The odds of being good enough to play professional football are infinitesimal, but Tiki has the right genes, which means that Ronde does too.

I believe in the power of the human spirit, but I also know the power of genetics. We are incredibly sophisticated machines. Our bodies and brains are the result of the orderly operation of tens of thousands of genes. These bodies and brains are capable of fantastic feats, and each is unique. Our genes and our environment have molded each of us into the one of a kind individuals that we are today. Unfortunately not all environments are equally good. Growing up with proper nourishment, free from toxins, with regular exercise will lead to a healthier adult body. Similarly, not all combinations of genes are equally good. A diverse complement of genes, high in heterozygosity, will provide a buffer against a fluctuating environment. That buffering effect allows some people to grow stronger, faster, healthier, smarter, and better looking.

If nothing else, the advantages of interracial people give us food for thought. Perhaps this knowledge can help break down some of the walls that still separate the human race. Interracial dating and even interracial fraternization are still frowned upon by many. It is my hope that this book leads to increased discussion and a reevaluation of these taboos.

In the early 1930s, Australian adventurer Michael "Mick" Leahy set out to explore some of the uncharted territory of New Guinea. He encountered tribespeople who had never met Europeans before. According to Mick, the locals were "utterly thunderstruck by our appearance. . . . One old chap came forward gingerly with open mouth, and touched me to see if I was real. Then he knelt down, and rubbed his hands over my bare legs, possibly to find if they were painted." Later interviews with village elders revealed that they were confused by the newcomers' white skin and by their weapons: "We thought the gun was just for shooting pigs and that it couldn't hurt men."

The natives weren't even sure that Leahy and his team were human. To find out they secretly watched the explorers to see if they had normal bodily functions. Finally a native scout returned with the report they'd been eagerly awaiting, "Their skin may be different, but their shit smells bad like ours."

That story always makes me laugh. I think it strikes just the right balance in pointing out how different we can be and still be the same. Human variation is real, and it's foolish to ignore it or sweep it under the rug. It's not something to be ashamed of or to avoid for fear of conflict. Clearly there are differences between us, but instead of making us incompatible they make us more compatible. Our diversity is a gift, and to keep ourselves separated and compartmentalized would be to waste that gift.

Notes

1. Your Wedding Night

Introduction

Pg #s

1 For a discussion of the universality of the incest taboo, see S. G. Frayser, *Varieties of Sexual Experience: An Anthropological Perspective on Human Sexuality* (New Haven: HRAF Press, 1985).

1 *The X-Files*. Episode no. 75, first broadcast October 11, 1996, by Fox. Directed by Kim Manners and written by Glen Morgan, James Wong

2 Morton, N. E. "Effect of Inbreeding on IQ and Mental Retardation." *Proceedings of the National Academy of Science* 75 (1978): 3906–8.

2 Dobzhansky, Theodosius. "Nothing in Biology Makes Sense Except in the Light of Evolution." *American Biology Teacher* 35 (1973):125–9.

The Evolution Will Not Be Televised

2 The "Scopes Monkey Trial" is now believed to have been arranged as a publicity stunt by the businessmen of Dayton, Tennessee, to attract attention to their small town. Scopes was in on the scheme. Their plan worked. *Inherit the Wind*, a stage play based on the trial, has been made into several film versions. However, the play was written in the 1950s as a thinly veiled attack on Senator McCarthy and contains some historical inaccuracies.

3 Schools in Cobb Country, Georgia, a suburb of Atlanta, placed stickers on their biology textbooks which read, "This textbook contains material on evolution. Evolution is a theory, not a fact, regarding the origin of living things. This material should be approached with an open mind, studied carefully and critically considered." In early

2005, a federal judge ruled that these stickers were unconstitutional because they send "a message that the school board agrees with the beliefs of Christian fundamentalists and creationists."

4 70 mph: "Cheetah." *World Almanac and Book of Facts* (Mahwah, NJ: Primedia, 1998).

4 My searches for "cheetah" and "sex" were performed on the Google Web search engine; however, I suspect that any Web search would yield a similar skew in favor of sex.

4 Evolved to avoid incest: There are, of course, documented cases of incest in certain isolated communities and royal families, but it's extremely rare, except in Greek tragedies.

Yet They Called Him the Georgia Peach

6 The Ty Cobb story comes from Al Stump, *Cobb* (Chapel Hill: Algonquin Books, 1996). The book is a fascinating read, mostly because Ty Cobb was so crazy. Al Stump ghostwrote Cobb's autobiography in 1960 and was privy to intimate details about his character. Stump writes, "I think, because he forced upon me a confession of his most private thoughts, along with the details of his life, that I know the answer to the central, overriding secret of his life. Was Ty Cobb psychotic throughout his baseball career? The answer is yes."

7 Paul Broca. On *the Phenomenon of Hybridity in the Genus Homo* (London: Longman, Green, Longman & Roberts, 1864).

7 Agassiz: E. C. Agassiz. *Louis Agassiz: His Life and Correspondence* (Boston: Houghton, Mifflin, 1895).

8 Distorted the data: As discussed extensively in Stephen Jay Gould, *The Mismeasure of Man* (New York: W. W. Norton, 1996).

8 Reference to the mule: According to *The American Heritage Dictionary of the English Language,* Fourth Edition.
   mu·lat·to (mŏŏ-lăt′ō, -lä′tō, myŏŏ-)
   n. pl. **mu·lat·tos** or **mu·lat·toes**
   1. A person having one white and one Black parent.
   2. A person of mixed white and Black ancestry.
   [Spanish mulato, *small mule, person of mixed race, mulatto,* from mulo, *mule,* from Old Spanish, from Latin m lus.]

Breaking the Law

9 "Whites who broke the law . . .": Karen Alonso. *Loving V. Virginia: Interracial Marriage* (Berkeley Heights, NJ: Enslow, 2000).

Pg #s

9 *Scott* v. *State* (1869), 39 Ga. 321, 324: "The amalgamation of the races is not only unnatural, but is always productive of deplorable results. Our daily observation shows us, that the offspring of these unnatural connections are generally sickly and effeminate, and that they are inferior in physical development and strength, to the full blood of either race."

9 "Were the Chinese to amalgamate . . .": Ronald Takaki. *Strangers from a Different Shore: A History of Asian Americans* (New York: Penguin, 1989).

9–10 The story of the Loving case: Karen Alonso. *Loving V. Virginia: Interracial Marriage* (Berkeley Heights, NJ: Enslow, 2000).

10 An abridged version of the Loving decision: http://www.ameasite.org/loving.asp.

Technicolor Lines

11 The Spike Lee quote is from Larry Elder, "Doing the Wrong Thing, Again," *FrontPage Magazine*, June 8, 1999, http://www.front-pagemag.com/Articles/ReadArticle.asp?ID=2826.

12 Julia Roberts on Denzel Washington: Allison Samuels. "Will It Be Denzel's Day?" *Newsweek*, February 25, 2002.

We Dare Maintain Our Rights
(Alabama State Motto)

13 "Alabama was the last state . . .": Article IV, Section 102 of Alabama's 1901 Constitution decrees: "The legislature shall never pass any law to authorize or legalize any marriage between any white person and a negro, or descendant of a negro."

13 This story was widely covered, here's just one article: Leonard Greene, "Intermarriage Ban Targeted in Alabama." *New York Post*, November 7, 2000.

13 The Hulond Humphries saga: Russ Jamieson, "Alabama Town Fears New School Superintendent's Alleged Bigotry," *CNN*, July 2, 1997, http://www.cnn.com/US/9707/02/humphries.return/.

14 Miscegenation hoax in 1864 election: Sidney Kaplan. "The Miscegenation Issue in the Election of 1864." *Journal of Negro History* 34, no 3 (1949): 274–343.

14 James Weldon Johnson was a poet, songwriter, and oddly, the U.S. consul to Venezuela in 1906. His song "Lift Ev'ry Voice and Sing" became known as the "Negro National Anthem."

14 "Open the bedroom doors . . .": James T. Patterson. *Brown v. Board of Education: A Civil Rights Milestone and Its Troubled Legacy* (London: Oxford University Press, 2002).

14–15 "Going over Niagara Falls . . .": Herbert Ravenel Sass. "Mixed Schools and Mixed Blood." *Atlantic Monthly* 198, November 1956.

15 "We must make the federal government . . .": President Harry Truman, addressing the annual convention of the NAACP, June 1947.

15 "I hope not . . .": *Chicago Daily Defender*, September 12, 1963.

15 "More likely than ever to live in racially integrated neighborhoods . . .": Susan Welch et al. *Race and Place: Race Relations in an American City* (London: Cambridge University Press, 2001).

15 "6 percent of married couples . . .": The number is 5.7 percent if you only look at race, but 7.4 percent if you include couples where one spouse is Hispanic and the other is not. For a discussion of why the 2000 Census grouped Hispanic origin separately from race, see Chapter Seven. From the U.S. Census Report, "Married-Couple and Unmarried-Partner Households: 2000," http://www.census.gov/prod/2003pubs/censr-5.pdf.

15–16 "90% correlation rate . . .": Jared Diamond. *The Third Chimpanzee: The Evolution and Future of the Human Animal* (New York: HarperCollins, 1992).

Some of My Best Friends Are White People (but I Wouldn't Want My Daughter to Marry One)

16 "86 percent of blacks thought . . .": Darryl Fears and Claudia Deane. "Biracial Couples Report Tolerance." *Washington Post*, July 5, 2001.

2. THE LEANING TOWER

Introduction

21 Check out the official Web site for The Leaning Tower of Pisa: http://torre.duomo.pisa.it/.

22 "Barbra Streisand, for example . . .": Anne Fadiman, "Barbra—Puts Her Career on the Line With 'Yentl'—And Learns New Lessons about Her Power and Her Femininity." *Life*, December 1983.

Symmetry and Health

Pg #s

25 Reviews of symmetry and health: Thornhill, R., and A.P. Moller. "Developmental Stability, Disease and Medicine." *Biological Reviews of the Cambridge Philosophical Society* 72 (1947): 497–548; Palmer, A.R., and C. Strobeck. "Fluctuating Asymmetry: Measurement, Analysis, Patterns." *Annual Review of Ecology and Systematics* 17 (1986): 391–421.

26 Forest tent caterpillar moth symmetry: Naugler, C.T., and S. M. Leech. "Fluctuating Asymmetry and Survival Ability in the Forest Tent Caterpillar Moth Malacosoma disstri: Implications for Pest Management." *Entomologia Experimentalis et Applicata* 70 (1994): 295–8.

26 Symmetrical flowers: A. P. Moller. "Bumblebee Preference for Symmetrical Flowers." *Proceedings of the National Academy of Science* 92 (1995): 2288–92.

26 Rural belize: Waynforth, D. "Fluctuating Asymmetry and Human Male Life History Traits in Rural Belize." *Proceedings of the Royal Society of London Series B* 265 (1999): 1497–501.

26 Symmetrical horses: Manning, J. T. and L. Ockenden. "Fluctuating Asymmetry in Race Horses." *Nature* 370 (1994): 185–6.

26 Faster running times: Manning, J. T., and L. J. Pickup. "Symmetry and Performance in Middle Distance runners." *International Journal of Sports Medicine* 19 (1998): 205–9.

26 "Sperm actually swim faster . . .": Manning, J. T., D. Scutt, and D. I. Lewis-Jones. "Developmental Stability, Ejaculate Size, and Sperm Quality in Men." *Evolution and Human Behaviour* 19 (1998): 273–82.

26 "Women with More Symmetrical Breasts . . .": Møller, A. P., M. Soler, and R. Thornhill. "Breast Asymmetry, Sexual Selection, and Human Reproductive Success." *Ethology and Sociobiology* 16 (1995): 207–19.

26 "Symmetrical people are bigger . . .": Manning, J. T. "Fluctuating Asymmetry and Body Weight in Men and Women: Implications for Sexual Selection." *Ethology and Sociobiology* 16 (1995): 145–53.

26 "More likely to suffer from depression . . .": Martin, S., J. T. Manning, and C. D. Dowrick. "Fluctuating Asymmetry, Relative Digit Length and Depression in Men." *Evolution and Human Behavior* 20 (1999): 203–14.

26 "Higher on intelligence tests . . .": Blinkhorn, S. "Symmetry as Destiny—Taking a Balanced View of IQ." *Nature* 387 (1997): 849–50.

<u>Pg #s</u>

26–27 "More likely to respond aggressively . . .": Benderlioglu, Z., P. W. Sci-
ulli, and R. J. Nelson. "Fluctuating Asymmetry Predicts Human
Reactive Aggression." *American Journal of Human Biology* 16
(2004): 458–69.

27 "When barn swallows are infected: Moller, A. P. "Parasites Dif-
ferentially Increase the Degree of Fluctuating Asymmetry in
Secondary Sexual Characters." *Journal of Evolutionary Biology*
5 (1992): 691–9.

27 "More symmetrical flies are more likely . . .": Moller, A. P. "Sexual
Selection, Viability Selection and Developmental Stability in the
Domestic Fly, Musca domestica." *Evolution* 50 (1995): 746–52.

28 For a fascinating look at the sex lives of giant water bugs, see Adrian
Forsyth, *A Natural History of Sex: The Ecology and Evolution of
Mating Behavior* (New York: Scribner's, 1986).

Hey Baby, What's Your Symmetry?

29 "Significant relationship between symmetry and attractiveness . .
.": Møller, A., and R. Thornhill. "Developmental Stability and
Sexual Selection: A Meta-analysis." *American Naturalist* 151
(1998): 174–92.

29 "Symmetrical faces more attractive . . .": Grammar, K., and R. Thorn-
hill. "Human (*Homo sapiens*) Facial Attractiveness and Sexual
Selection: The Role of Symmetry and Averageness." *Journal of
Comparative Psychology* 108 (1994): 233–42, and Perrett, D. et al.
"Symmetry and Human Facial Attractiveness." *Evolution and
Human Behavior* 20 (1999): 295–307.

29 "Symmetrical men lose their virginity 3–4 years earlier . . .": Thornhill,
R. and S. W. Gangestad. "Human Fluctuating Asymmetry and
Sexual Behavior." *Physiological Science* 5 (1994): 297–302.

30 "Women with more symmetrical partners were much more likely
to orgasm . . .": Thornhill, R., S. W. Gangestad, and R. Comer.
"Human female Orgasm and Mate Fluctuating Asymmetry."
Animal Behavior 50 (1995): 1601–15.

31 Upsuck theory: Fox, C. A. "Uterine Sucking During Orgasm." *Brit.
Med. J.* 1 (1967): 300; and Fox, C. A. et al. "Measurement of Intra-
Vaginal and Intra-Uterine Pressure During Human Coitus by
Radio-Telemetry." *Journal of Reproductive Fertility* 22 (1970):
243–51.

32 "Female Japanese scorpionflies are more attracted to . . .": Thornhill,
R. "Female Preference of Males with Low Fluctuating Asymmetry in the Japanese Scorpionfly." *Behavioral Ecology* 3 (1992):
277–83.

Pg #s

33–34 "Pick out symmetrical, high-quality males based on smell . . .":
Gangestad, S. W., and R. Thornhill. "Menstrual Cycle Variation
in Women's Preferences for the Scent of Symmetrical Men." *Proceedings of the Royal Society of London*, B, 265 (1998): 927–33.

Symmetrical Men and the Women Who Love Them

36–37 "Casual sex with no strings attached . . .": Clark, R. D. and E. Hatfield.
"Gender Differences in Receptivity to Sexual Offers." *Journal of
Psychology & Human Sexuality* 2 (1989): 39–55.

   38 "Giant Water Bugs' Gender Roles are the Opposite of Humans.": For
a fascinating look at the sex lives of giant water bugs, see Adrian
Forsyth, *A Natural History of Sex: The Ecology and Evolution of
Mating Behavior* (New York: Scribner's, 1986).

A non sequitur from the tastes-like-chicken department: According
to renowned wildlife biologist W. S. Bristowe, the giant water bug
tastes like gorgonzola. I must confess that this fact, while interesting, has affected my ability to enjoy certain pasta dishes.

In a 1953 article called, "Insects as Food," Bristowe documented some
of his culinary adventures. He had this to say about the giant
water bug: "The strongest flavour I experienced was provided by
giant water-bugs about 3 in. in length (Lethocerus indicus Lep.
and Sev.) whose relatives in Mexico are also eaten. The Lethocerus
is caught in nets and at different seasons was fetching from 5 to
20 satangs (1d. to 4d.) each in Bangkok."

It reached the tables of princes as well as peasants. The usual methods
of preparing it for table are as follows: 1) Steam thoroughly and
then soak in shrimp sauce. The flavour reminded me of Gorgonzola cheese. 2) After cooking, pound up and use for flavouring
sauces or curries. A popular sauce called namphla is made by
mixing shrimps, lime juice, garlic, and pepper and then adding
the pounded-up bugs.

Questions I'm Not Going to Answer

39–40 "Clear, unblemished skin was historically associated with a healthy
individual . . .": Gangestad, S. W. and D. M. Buss. "Pathogen
Prevalence and Human Mate Preferences." *Ethology and Sociobiology* 14 (1993): 89–96.

3. BUILDING THE TOWER

Introduction

Pg #s

43 For more information about Easter Island and modern attempts to replicate the construction of the *moai*, I recommend an episode of NOVA titled, "Secrets of Lost Empires: Easter Island." You can also check out the Web site: http://www.pbs.org/wgbh/nova/easter/.

Lazy Carpenters and Rotten Wood

46–47 For more information about Gaucher disease, see the National Gaucher Foundation home page: http://www.gaucherdisease.org/.

48 For more information about Huntingon's, see the Huntington's Disease Society of America: http://www.hdsa.org/.

48–49 For more information about sickle cell disease, see the Sickle Cell Disease Association of America: http://www.sicklecelldisease.org/.

49 For more information about cystic fibrosis see the Cystic Fibrosis Foundation: http://www.cff.org/.

49–50 Cystic fibrosis and resistance to cholera: Sereth H., T. Shoshani, N. Bashan, and B. S. Kerem. "Extended Haplotype Analysis of Cystic Fibrosis Mutations and Its Implications for the Selective Advantage Hypothesis." *Human Genetics* 92 (1993): 289–95.

Gabriel S. E, et al. "Cystic Fibrosis Heterozygote Resistance to Cholera Toxin in the Cystic Fibrosis Mouse." *Science* 266 (1994): 107–9.

Gabriel S. University of North Carolina at Chapel Hill 1994 report. *Discover* (March 1995).

Others think that the real advantage of CF is resistance to tuberculosis: Meindl, R. S. "Hypothesis: A Selective Advantage for Cystic Fibrosis Heterozygotes." *American Journal of Physical Anthropology* 74 (1987): 39–45.

Others think it's influenza: Shier, W. T. "Increased Resistance to Influenza as a Possible Source of Heterozygote Advantage in Cystic Fibrosis." *Medical Hypotheses* 5 (1979): 661–7.

Still others think it's asthma: Schroeder, S. A, D. M. Gaughan, and M. Swift. "Protection Against Bronchial Asthma by CFTR Delta F508 Mutation: A Heterozygote Advantage in Cystic Fibrosis." *Nature Medicine* 1 (1995): 703–5.

A Quick Look at My Brain

Pg #s

50 "Recently Gregory Cochran . . .": "Natural genius?" *Economist*, June 2, 2005.

51 "Protection against tuberculosis . . .": Diamond, J. M.: "Tay-Sachs Carriers and Tuberculosis Resistance." (Letter) *Nature* 331 (1988): 666.

Heterozygosity and Symmetry

54–55 A thorough review of heterozygosity and symmetry is found in: Mitton, J. B., and M. C. Grant. "Associations among Protein Heterozygosity, Growth Rate, and Developmental Homeostasis." *Annual Review of Ecology and Systematics* 15 (1984): 479–99.

55 "Some Tamarins are more symmetrical . . .": Hutchison, D. W., and J. M. Cheverud. "Fluctuating Asymmetry in Tamarin (*Saguinus*) Cranial Morphology: Intra- and Interspecific Comparisons between Taxa with Varying Levels of Genetic Heterozygosity." *Journal of Heredity* 86 (1995): 280–8.

55 "A study of cheetah symmetry . . .": Wayne, R. K., W. S. Modi, and S. J. O'Brien. "Morphological Variability and Asymmetry in the Cheetah *Acinonyx jubatus*, a Genetically Uniform Species." *Evolution* 40 (1986): 78–85.

MHC Heterozygosity and Disease

57 "AIDS onset in patients with different levels of MHC heterozygosity . . .": Carrington, M. et al. "HLA and HIV-1: Heterozygote Advantage and B*35-Cw*04 Disadvantage." *Science* 283 (1999): 1748–52; and Tang, J. et al. "HLA Class I Homozygosity Accelerates Disease Progression in Human Immunodeficiency Virus Type 1 Infection." *AIDS Research and Human Retroviruses* 15 (1999): 317–24.

57 "A study of Gambians infected with Hepatitis B . . .": Thursz, M. R., et al. "Heterozygote Advantage for HLA Class-II Type in Hepatitis B Virus Infection." *Nature Genetics* 17 (1997): 11–12.

57–58 "Mice infected with fungus . . .": McClelland, E. E., D. L. Granger, and W. K. Potts. "Major Histocompatibility Complex-Dependent Susceptibility to *Cryptococcus neoformans* in Mice." *Infection and Immunity* 71 (2003): 4815–7.

"Mice simultaneously infected with Salmonella and a neurologically damaging virus . . .": McClelland, E. E., D. J. Penn, and W. K. Potts. "Major Histocompatibility Complex Heterozygote Superiority during Coinfection." *Infection and Immunity* 71 (2003): 2079–86.

Pg #s

"Mice simultaneously infected with three different strains of Salmo-nella and one of Listeria . . .": Penn, D. J., K. Damjanovich, and W. K. Potts. "MHC Heterozygosity Confers a Selective Advantage Against Multiple-Strain Infections." *Proceedings of the National Academy of Sciences* 99 (2002): 11260–4.

58–59 "Heterozygosity and growth has been observed in many species . . .": Mitton, J. B., and M. C. Grant. "Associations among Protein Heterozygosity, Growth Rate, and Developmental Homeostasis." *Annual Review of Ecology and Systematics* 15 (1984): 479–99.

Wham, Bam, Thank You Clam

61 "Being a highly heterozygous clam gives you a big advantage . . .": Brown, J. L. "The New Heterozygosity Theory of Mate Choice and the MHC." *Genetica* 104 (1999): 215–21.

61–62 "Macaque males with higher levels of MHC heterozygosity . . .": Sauermann, U. et al. "Increased Reproductive Success of MHC Class-II Heterozygous Males Among Free-Ranging Rhesus Macaques." *Human Genetics* 108 (2001): 249–54.

Shiny, Happy Butterflies

62 "Male sulfur butterflies that are heterozygous . . .": Brown, J. L. "The New Heterozygosity Theory of Mate Choice and the MHC." *Genetica* 104 (1999): 215–21.

63 "Heterozygous males had the longest throat feathers . . .": Aparicio, J. M., P. J. Cordero, and J. P. Veiga. "A Test of the Hypothesis of Mate Choice Based on Heterozygosity in the Spotless Starling." *Animal Behavior* 62 (2001): 1001–6.

63 "Minnows with high levels of heterozygosity . . .": Muller, G. and P. I. Ward. 1995. "Parasitism and Heterozygosity Influence the Secondary Sexual Characters of the European Minnow." *Ethology* 100 (1995): 309–19.

But What Do the T-shirts Say?

64 "Similar MHC types preferred the same perfumes . . .": Milinski, M., and C. Wedekind. "Evidence for MHC-Correlated Perfume Preferences in Humans." *Behavioral Ecology* 12 (2001): 140–9.

64–65 "Another intriguing T-shirt sniffing experiment . . .": Thornhill, R. et al. "Major Histocompatibility Complex Genes, Symmetry, and Body Scent Attractiveness in Men and Women." *Behavioral Ecology* 14 (2003): 668–78.

4. Assembling the Crew

Introduction

Pg #s

68 For a detailed account of bedbugs' "traumatic copulation" and other fascinating sexual antics of the animal kingdom, see Adrian Forsyth, *A Natural History of Sex: The Ecology and Evolution of Mating Behavior* (New York: Scribner's, 1986).

Yes, We Have No Bananas

72–73 Banana facing extinction . . . again . . . : Dan Koeppel. "Can this Fruit Be Saved?" *Popular Science* (August 2005).

But I Heard That Race Doesn't Exist

75 Darwin blamed his incestuous marriage for his sick children: A. Desmond and J. Moore, *Darwin* (London: Michael Joseph, 1991).

We Marry Our Own

76 "All Icelanders are related to each other . . .": Bouchard, C. "Genetic Basis of Racial Differences," *Canadian Journal of Sports Science* 13 (1988): 104–8.

77 "One important Jewish tradition is that of the kohanim . . .": Thomas, M. et al. "Origins of Old Testament Priests" *Nature* 394 (1998): 138–40.

78 "Less than 1 percent of Jewish women had children with non-Jewish men . . .": Nebel, A. et al. "The Y Chromosome Pool of Jews as Part of the Genetic Landscape of the Middle East." *American Journal of Human Genetics* 69 (2001): 1095–112.

Mixing It Up

79 For a discussion of the history of human populations, including the Middle East as a crossroads, see Steve Olson, *Mapping Human History: Genes, Race, and Our Common Origins* (New York: Houghton Mifflin, 2002).

80 "6.8 to 22.5 percent of African American genes to be of European origin . . .": Parra, E. et al. "Estimating African American Admixture Proportions by Use of Population-Specific Alleles." *American Journal of Human Genetics* 63 (1998): 1839–51, and Chakraborty, R. et al. "Caucasian Genes in American Blacks: New Data." *American Journal of Human Genetics* 50 (1992): 145–55.

You're What?

Pg #s

80–81 "Geneticists tried to solve the riddle of the origin of the Lemba . . .":
Thomas, M. et al. "Y Chromosomes Traveling South: The Cohen
Modal Haplotype and the Origins of the Lemba—The 'Black Jews
of Southern Africa.'" *American Journal of Human Genetics* 66
(2000): 674–86.

81 The fable of the Good Samaritan is told in the *Gospel of Luke*
(10:25–37).
"Samaritans are one of the most inbred . . .": Batsheva Bonne–Tamir.
"The Samaritans: A Living Ancient Isolate." In *Population Struc-
ture and Genetic Disorders*, edited by Aldur Eriksson (London:
Academic Press, 1980).

82 "46 percent of couples were first or second cousins . . .": Steve Sailer.
"Cousin Marriage Conundrum." *American Conservative*, January 13, 2003.

The Emperor Has No Race

83 "FDA recently approved BiDil . . .": Susan Heavey. "Heart Drug
for U.S. Blacks Can Cut Costs—study." *Reuters*, December 12,
2005.

83 "95 percent of Asian Americans are lactose-intolerant . . .": Huang, S.
S., and T. M. Bayless. "Milk and Lactose Intolerance in Healthy
Orientals." *Science* 160 (1968): 83–4.

83–84 "Recent issue of Scientific American . . .": The article in question
is Michael J. Bamshad and Steve E. Olson, "Does Race Exist?"
Scientific American (December 2003).

85 DNAPrint Genomics: Check out their Web site at www.dnaprint.
com. The forensic service I mention is called DNAWitness™. In
an impressive display of politically correct doubletalk, the Web
site discusses DNAWitness at length without actually using the
word "race."

85 "A massive tome called . . .": Luca Cavalli-Sforza, L., Paolo Menozzi,
and Alberto Piazza. *The History and Geography of Human Genes*
(Princeton: Princeton University Press, 1994).

Opening Pandora's Box

88 "Are certain races better at certain things?": For an interesting dis-
cussion of this highly charged topic, I recommend Jon Entine,
*Taboo: Why Black Athletes Dominate Sports and Why We're
Afraid to Talk about It* (New York: PublicAffairs, 2000).

5. Inspecting the Towers

Introduction

Pg #s

92 "The inhabitants of Pitcairn found quite the opposite . . .": Harry Lionel Shapiro. *The Heritage of the Bounty* (Brooklyn, NY: AMS Press, 1979).

A Little Bit of Pop Culture

93 "In an emotional acceptance speech . . .": Berry won the Best Actress Academy Award for her role in *Monster's Ball.* You can watch her speech and read the transcript here: http://www.american-rhetoric.com/speeches/halleberryoscarspeech.htm.

93 "This caused a backlash . . .": Jack E. White. "I'm Just Who I Am." *Time*, May 5, 1997.

93–94 "He recently opened up . . .": Charles Barkley. *Who's Afraid of a Large Black Man?* (New York: Penguin, 2005).

Only Julia Knows for Sure

94–95 "Lyle Lovett was their poster boy for asymmetry . . .": Geoffrey Cowley. "The Biology of Beauty." *Newsweek*, June 3, 1996.

Milk and Turkeys

95–96 "Average American drank 34 gallons of milk . . .": Judy Putnam and Jane Allshouse. "Trends in U.S. Per Capita Consumption of Dairy Products, 1909 to 2001." *Amber Waves* (June 2003), http://www.ers.usda.gov/Amberwaves/June03/DataFeature/.

96 "Selectively breeding the cows that gave the most milk . . .": Roberts, M. "U.S. Animal Agriculture: Making the Case for Productivity." *AgBioForum* 3(2&3) (2000): 120–6, http://www.agbioforum.org/v3n23/v3n23a08–roberts.htm.

96 "Increase the size of the average turkey breast . . .": Mary Bergin. "Time to Talk Turkey," *Capital Times*, November 21, 2005, http://www.madison.com/tct/features/stories/index.php?ntid=62352&ntpid=2.

Hybrid Vigor

98 "In 1933, hybrid corn only accounted for about 1 percent . . .": G. F. Sprague. "Development and Adoption of Hybrid Corn." *Illinois Research* (spring/summer 1992), http://web.aces.uiuc.edu/vista/pdf_pubs/IRSpSm92.PDF.

Pg #s

99 "Wheat, cotton, and alfalfa have made significant gains . . .": Richard Manning. *Against the Grain: How Agriculture Has Hijacked Civilization* (New York: North Point Press, 2004).

99 "Calf crop increase . . .": Robert Wells. "Heterosis . . . Hype or Legit?" *Livestock* (October 2005), http://www.noble. org/Ag/Livestock/Heterosis/.

99 "Crossing dwarf wheat . . .": Richard Manning. *Against the Grain: How Agriculture Has Hijacked Civilization* (New York: North Point Press, 2004).

Rin Tin Tin's Granddad

99–100 "Horand v Grafeth . . .": http://www.germanshepherds. com/thegsd/history/.

100 "Suffer from many genetic problems . . .": George A. Padgett. *Control of Canine Genetic Diseases* (New York: Howell Book House, 1998).

100 "Poodles are at risk for . . .": http://www.poodleclubofamerica.org/health.htm.

What the Hell Is a Zonkey?

102 "Comes from the Spanish word . . .": According to *The American Heritage Dictionary of the English Language, IV Edition*:
mu·lat·to (mo͞o-lăt´ō, -lä´tō, myo͞o-)
n. pl. **mu·lat·tos** or **mu·lat·toes**
1. A person having one white and one Black parent.
2. A person of mixed white and Black ancestry.
[Spanish mulato, *small mule, person of mixed race, mulatto*, from mulo, *mule*, from Old Spanish, from Latin m lus.]

102 "According to the American Donkey and Mule Society . . .": http:// www.lovelongears.com/.

103 When asked what a liger is, Napoleon Dynamite says, "It's pretty much my favorite animal. It's like a lion and a tiger mixed . . . bred for its skills in magic."

103 "A viable sheep/goat hybrid . . .": Jonathan Amos, "'Funny Creature Toast of Botswana." *BBC News Online*, July 3, 2000, http://news. bbc.co.uk/1/hi/sci/tech/813466.stm.

Back to the Wild

103–104 "Gazelles were transplanted . . .": Alados, C. L., J. Escos, and J. M. Emien. "Fluctuating Asymmetry and Fractal Dimension of the Sagittal Suture as Indicators of Inbreeding Depression in Dama and Dorcas Gazelles." *Canadian Journal of Zoology* 73 (1995): 1967–74.

Pg #s

105 "Compare skulls of pre–bottleneck seals . . .": Hoelzel, A. R. et al. "Impact of a Population Bottleneck on Symmetry and Genetic Diversity in the Northern Elephant Seal." *Journal of Evolutionary Biology* 15 (2002): 567–75.

105 "Supervolcano in Indonesia . . .": Some have proposed that the Toba caldera in Sumatra erupted about seventy thousand years ago with a force three thousand times that of Mount St. Helens. The ash spewed from this supervolcano plunged the earth into a volcanic winter, decimating the human population. Ambrose, S. H. "Late Pleistocene Human Population Bottlenecks, Volcanic Winter, and Differentiation of Modern Humans." *Journal of Human Evolution* 34 (1998): 623–51.

105 "Small group of Vikings colonized Iceland . . .": Bouchard, C. "Genetic Basis of Racial Differences." *Canadian Journal of Sports Science* 13 (1988): 104–8.

105 "Inbred strains of house mice . . .": Leamy, L. "Morphometric Studies in Inbred and Hybrid House Mice. VII. Heterosis in Fluctuating Asymmetry at Different Ages." *Acta Zoologica Fennica* 191 (1992): 111–9.

105–106 "Selectively abort the self-pollinated . . .": Adrian Forsyth. *A Natural History of Sex: The Ecology and Evolution of Mating Behavior* (New York: Scribner's, 1986).

106 Charles Darwin in "On the Various Contrivances by which British and Foreign Orchids Are Fertilised by Insects, and on the Good Effects of Intercrossing"(1862).

Human Hybrid Vigor

106 "A 1957 study . . .": Hulse, F. S. "Exogamy and Heterosis." *Yearbook of Physical Anthropology* 9 (1957): 240–57.

107 "Study was carried out in Poland in 1970 . . .": Wolanski, N., E. Jarosz, and M. Pyzuk. "Heterosis in Man: Growth in Offspring and Distance Between Parents' Birthplaces." *Social Biology* 17 (1970): 1–16.

107–108 "Study in Hawaii on the genetics of intelligence . . .": Nagoshi, C. T., and R. C. Johnson. "The Ubiquity of G." *Personality and Individual Differences* 7 (1986): 201–7.

You Really Are Smarter than Your Parents

Pg #s

108 "Average IQ score has consistently been creeping up . . .": Mingroni, M. A. "The Secular Rise in IQ: Giving Heterosis a Closer Look." *Intelligence* 32 (2004): 65–83.

Frankenstein vs. Stalin?

110 "Weird sense of humor . . .": "Voting for Frankenstein." *BBC News Online*, February 26, 2003, http://news.bbc.co.uk/2/hi/south_asia/2800585.stm.
110–111 "Sometimes the Khasi do intermarry . . .": Khongsdier, R., and N. Mukherjee. "Effects of Heterosis on Growth in Height and Its Segments: A Cross–Sectional Study of the Khasi Girls in Northeast India." *Annals of Human Biology* 30 (2003): 605–21.

I ♥ Symmetry

112–113 "A groundbreaking study at UCLA . . .": Phelan, J. P., personal communication, October 1, 2005.

6. WHY DO FOOLS FALL IN LOVE?

Introduction

115 "In January of 1936 . . .": "On This Day," BBC News Online, April 29, 1986, http://news.bbc.co.uk/onthisday/hi/dates/stories/april/29/newsid_2500000/2500427.stm.

As with any political story, the abdication crisis has its share of conspiracy theories. In a decidedly less romantic turn, some think it was precipitated by Wallis Simpson's suspected Nazi ties: Emma Simpson, "Simpson's 'Nazi Past' Led to Abdication," January 9, 2003, http://news.bbc.co.uk/1/hi/uk/2644123.stm.

More than Skin Deep

116 "British, Chinese, and Indian women rate photos of Greek men . . .": Geoffrey Cowley, "The Biology of Beauty." *Newsweek*, June 3, 1996.
117 "Studies on the Ache and Hiwi tribes . . .": Jones, D., and K. Hill. "Criteria of Facial Attractiveness in Five Populations." *Human Nature* 4 (1993): 271–95.

Pg #s

117 "3 and 6 month old babies . . .": Langlois, J. H., L. A. Roggman, and L. A. Reiser–Danner. "Infants' Differential Social Responses to Attractive and Unattractive Faces." *Developmental Psychology* 26 (1990): 153–9.

So What Are the Rules?

118 "Clear skin is universally admired . . .": Gangestad, S. W. and D. M. Buss. "Pathogen Prevalence and Human Mate Preferences." *Ethology and Sociobiology* 14 (1993): 89–96.

Like Sand through the Hourglass

118 "Waist–to–hip ratio of .7 . . .": D. Singh. "Adaptive Significance of Waist-to-Hip Ratio and Female Physical Attractiveness." *Journal of Personality and Social Psychology* 65 (1993): 293–307.

118 "Miss America winners . . .": A. Mazur. "U.S. Trends in Feminine Beauty and Over Adaptation." *Journal of Sex Research* 22 (1986): 281–303.

119 ".7 waist–to–hip ratio are more fertile . . .": Zaadstra, B. M. et al. "Fat and Female Fecundity: Prospective Study of Effect of Fat Distribution on Conception Rates." *British Medical Journal* 306 (1993): 484–7.

119 "Full lips, a small jaw, and delicate chin . . .": Geoffrey Cowley. "The Biology of Beauty." *Newsweek*, June 3, 1996.

119 "Strong jaw and thick brow . . .": Geoffrey Cowley. "The Biology of Beauty." *Newsweek*, June 3, 1996.

Oh Brother

Pg #s

120 "Creepy in any culture . . .": S. G. Frayser *Varieties of Sexual Experience: An Anthropological Perspective on Human Sexuality* (New Haven: HRAF Press, 1985).

120 "All sorts of health problems . . .": Morton, N. E. "Effect of Inbreeding on IQ and Mental Retardation." *Proceedings of the National Academy of Science* 75 (1978): 3906–8.

120–121 "None married a person they had been raised with . . .": Talmon, Y. "Mate Selection in Collective Settlements." *American Sociological Review* 29 (1964): 491–508.

121 Shim-pua: Arthur P. Wolf. *Sexual Attraction and Childhood Association: A Chinese Brief for Edward Westermarck* (Palo Alto, CA: Stanford University Press, 1995).

Of Mice and Hutterites

Pg #s

121 "Mice prefer the scent . . .": Potts, W. K., C. J. Manning, and E. K. Wakeland. "Mating Patterns in Seminatural Populations of Mice Influenced by MHC Genotype." *Nature* 352 (1991): 619–21.

122 "Another T-shirt sniffing experiment . . .": Wedekind, C., and S. Furi. "Body Odour Preferences in Men and Women: Do They Aim for Specific MHC Combinations or Simply Heterozygosity?" *Proceedings of the Royal Society of London* B 264 (1997): 1471–9.

123 "A study of 411 Hutterite couples . . .": Ober, C. et al. "HLA and Mate Choice in Humans." *American Journal of Human Genetics* 61 (1997): 497–504.

123 "A separate study found that Hutterite couples . . .": Ober, C. "HLA and Reproduction: Lessons from Studies in the Hutterites." *Placenta* 16 (1995): 569–77.

Making Choices the Morning After

123 "Female sand lizards and adders . . .": Olsson, M. et al. "Sperm selection by females." *Nature* 383 (1996): 585.

124 "Bird studies have found . . .": Hasselquist, D., S. Bensch, and T. von Schantz. "Correlation between Male Song Repertoire, Extra-pair Paternity and Offspring Survival in the Great Reed Warbler." *Nature* 381 (1996): 229–32; and Kempenaers, B., F. Andriaesen, V. A. J. Noordwijk, and A. A. Dhondt. "Genetic Similarity, Inbreeding and Hatching Failure in Blue Tits: Are Unhatched Eggs Infertile?" *Proceedings of the Royal Society of London* 263 (1996): 179–85.

124–5 "Couples who have recurrent miscarriages . . .": Hedrick, P.W. "HLA–Sharing, Recurrent Spontaneous Abortion, and the Genetic Hypothesis." *Genetics* 119 (1998): 199–204.

Like Marries Like

125–6 "Not just long-term partners that look alike . . .": Griffiths, R. W., and P. R. Kunz. "Assortative Mating: A Study of Physiognomic Homogamy." *Social Biology* 20 (1973): 448–53.

126 "Researchers raised Japanese quail chicks . . .": Bateson, P. "Preferences for Cousins in Japanese Quail." *Nature* 295 (1982): 236–37; and Bateson, P. "Preferences for Close Relations in Japanese Quail." In *Acta XIX Congressus Internationalis Ornithologici*, Vol. I, edited by H. Ouellet (Ottawa: University of Ottawa Press, 1988).

Pg #s

126-7 "Colorful snow geese . . .": Cooke, F. and C. M. McNally. "Mate Selection and Colour Preferences in Lesser Snow Geese." *Behaviour* 53 (1975): 151-70; and Cooke, F. et al. "Assortative Mating in Lesser Snow Geese (*Anser caerulescens*)." *Behavior Genetics* 6 (1976): 127-40.

Oedipus Rats

127 "Male rats were raised by lemon-scented mothers . . .": Fillion, T. J., and E. M. Blass. "Infantile experience with Suckling Odors Determines Adult Sexual Behavior in Male Rats." *Science* 231 (1986): 729-31.

128 "Like the old song . . .": "I Want A Girl," words by Will Dillon, music by Harry Von Tilzer (1911).

In the Meme of the Father

132 Memes: Richard Dawkins. *The Selfish Gene* (Oxford: Oxford University Press, 1976).

133 "First commandment . . .": Exodus 20:3-17.

133 Doubting Thomas: John 20:24-29.

134 "'Judah hath dealt treacherously . . . '": Malachi 2:11-12.

134 "When King Ahab married her . . .": 1 Kings 16:29-33.

134 "Elijah later prophesizes . . .": 1 Kings 21:23.

Keeping It in the Family

135 "Between a man and his father's brother's daughter . . .": Bill, James A., and Robert Springborg. *Politics in the Middle East* (New York: HarperCollins College Publishers, 1994).

135-6 "Understanding the politics of the Middle East . . .": Steve Sailer. "Cousin Marriage Conundrum." *American Conservative*, January 13, 2003.

What's Next?

136 "Remember tamagotchis?": If you don't remember tamagotchis, they were egg-shaped, keychain sized virtual pets that were sold by Bandai. The name may come from the Japanese words for egg (*tamago*) and friend (*tomodachi*). These egg-friends were briefly the must-have toy of 1997.

136-7 "Published a study on race relations . . .": Gunnar Myrdal. *An American Dilemma: The Negro Problem and Modern Democracy* (New York: Harper & Row, 1944).

137 "A survey in 1958 . . .": David Greenberg. "White Weddings." *Slate*, June 15, 1999, http://www.slate.com/id/30352/.

7. Is the Pot Melting?

Introduction

140 "In 1963, then liberal writer . . .": Podhoretz, Norman. "My Negro Problem—And Ours." *Commentary* (1963): 35
140 "'The best hope for the future of American race relations.'": Besharov, D. J., and T. S. Sullivan. "One Flesh." *New Democrat* 8 no. 4 (1996): 19–21.

I Have a Dream

140–141 "One of the most famous speeches in American history . . .": Martin Luther King Jr. delivered this speech on August 28, 1963, at the Lincoln Memorial in Washington, D.C. You can read the text and listen to the speech here: http://www.americanrhetoric. com/speeches/Ihaveadream.htm.
141–2 "Brazilian Institute of Geography and Statistics . . .": Levine, Robert M. and John J. Crocitti, ed. *The Brazil Reader: History, Culture, Politics* (Durham, Duke University Press, 1999).

Welcome to the Hotel California

142–3 "I looked at a list of the metropolitan areas . . .": All of the U.S. Census data that I cite is available at http://www.census.gov. However, I recommend starting your search at http://www. censusscope.org/, which presents the 2000 U.S. Census data in more user-friendly charts, graphs, and maps.
143–4 "The Public Policy Institute of California analyzed birth records . . .": Tafoya, S. "Check One or More . . . Mixed Race and Ethnicity in California." *California Counts: Population Trends and Profiles* 1 (2000): 1–11, http://www.ppic.org/content/pubs/CC_100STCC. pdf.

What about the Rest of the Country?

145 "Six general choices for race . . .": To see a snapshot of the actual 2000 U.S. Census question on race go to http://www.censusscope. org/us/chart_multi.html and click on "race."
145 "'Hispanics and Latinos may be of any race.'": http://www.census. gov/population/www/socdemo/race/racefactcb.html.

<u>Pg #s</u>

146 "Census Bureau added a separate question concerning the respon-dent's 'origin . . . '": To see a snapshot of the actual 2000 U.S. Census question on origin go to http://www.censusscope.org/us/chart_race.html and click on "Hispanic ethnicity."

146 "Unfortunately this seems to have led to some confusion . . .": http://www.censusscope.org/us/map_multiracial.html.

The Great Divide

147 "Color coded map of America . . .": You can see a color coded map depicting the distribution of the American multiracial population here: http://www.censusscope.org/us/map_multiracial.html.

147–8 "Of the top ten most interracial states . . .": You can see all fifty states and the District of Columbia ranked by how multiracial their population is here: http://www.censusscope.org/us/rank_multi.html.

148 "The West has a far larger Hispanic population than the East.": http://www.censusscope.org/us/map_hispanicpop.html.

148 "12.31 percent of Illinois' population is Hispanic . . .": http://www.censusscope.org/us/rank_race_hispanicorlatino_alone.html.

148 "Top five fastest growing states from 1990 to 2000 were all western . . .": http://www.censusscope.org/us/rank_popl_growth.html.

Who's Mixing and Who Isn't

149 "We can calculate what's called a Dissimilarity Index . . .": http://www.censusscope.org/segregation.html.

149 "Fifty American cities that are most segregated with respect to whites and blacks . . .": http://www.censusscope.org/us/rank_dissimilarity_white_black.html.

149 "Fifty most segregated cities with respect to whites and Asians . . .": http://www.censusscope.org/us/rank_dissimilarity_white_asian.html.

149–50 "14 percent of respondents who checked Asian also checked another box . . .": http://www.aecf.org/kidscount/categories/counts.htm.

150 "Two-thirds of Asians in their twenties born in the U.S. married outside their race . . .": Nancy Etcoff. *Survival of the Prettiest: The Science of Beauty* (New York: Anchor Books, 2000).

150 "The reverse is true for Asians in this country . . .": Steve Sailer. "Is Love Colorblind?" *National Review*, June 14, 1997.

Get Ready for a Mixed Race Explosion

151 "White majority is less of a majority than ever before . . .": http://www.censusscope.org/us/chart_race.html.

Pg #s

151 "Its population increased by 19.7 percent in the 1990s . . .": Perry, Marc J. and Paul J. Mackun. "Population Change and Distribution: 1990–2000." *Census 2000 Brief* (April 2001), http://www.census.gov/prod/2001pubs/c2kbr01–2.pdf.

152 "Only 1.9 percent of adults were interracial . . .": http://www.aecf.org/kidscount/categories/counts.htm.

Ghost in the Gene Machine

155 "An obsessive student of batting . . .": You can read about Ted Williams' intense training in Ted Williams, *My Turn at Bat : The Story of My Life* (Lehi, UT: Fireside Sports Classics, 1988). Here's an excerpt: http://www.baseballlibrary.com/baseballlibrary/excerpts/ted_williams3.stm.

156 "The future astronaut described him as . . .": "Significant moments in sports and war." *ESPN Page 2*, http://ad.abctv.com/page2/s/list/warandsports.html.

156 "His eyesight was measured as 20/10 . . .": "In Every Sense, Williams Saw More Than Most." *USA Today*, June 6, 1996, http://www.usatoday.com/sports/baseball/sbbw0725.htm.

157 "Australian adventurer Michael Mick Leahy set out to explore . . .": Bob Connolly and Robin Anderson. *First Contact: New Guinea's Highlanders Encounter the Outside World* (New York: Viking, 1987).

Index